PHILOSOPHICAL PAPERS

VIENNA CIRCLE COLLECTION

VOLUME 8

EDITOR: BRIAN McGUINNESS

FRIEDRICH WAISMANN (1896-1959)

FRIEDRICH WAISMANN

PHILOSOPHICAL PAPERS

Edited by

BRIAN McGUINNESS

With an Introduction by

ANTHONY QUINTON

D. REIDEL PUBLISHING COMPANY

DORDRECHT-HOLLAND/BOSTON-U.S.A.

Library of Congress Cataloging in Publication Data

Waismann, Friedrich.
    Philosophical papers.

        (Vienna circle collection)
        'Bibliography of works by Friedrich Waismann': p.
        Includes index.
        CONTENTS: Editor's note. — The nature of the axiom
    of reducibility. — A logical analysis of the concept of
    probability [etc.]
            1. Philosophy—Collected works. I. Title.
    II. Series.
    B1669.W41 1976   100   76-41280
    ISBN 90-277-0712-X
    ISBN 90-277-0713-8 pbk.

Chapters I, II, III, V, VI, and VIII translated from the German by Hans Kaal
Chapter IV translated from the Dutch by Arnold Burms and Philippe van Parys.

Published by D. Reidel Publishing Company,
P.O. Box 17, Dordrecht, Holland

Sold and distributed in the U.S.A., Canada, and Mexico
by D. Reidel Publishing Company, Inc.
Lincoln Building, 160 Old Derby Street, Hingham,
Mass. 02043, U.S.A.

CONTENTS

INTRODUCTION

Friedrich Waismann was born in Vienna in 1896 and lived there until the time of the *Anschluss* in 1938. From then until his death in 1959 he lived in England; this, apart from a brief period at Cambridge early on, was almost wholly at Oxford, where he held the posts, first, or reader in the philosophy of mathematics and then of reader in the philosophy of science. He was of Jewish descent – his father being Russian, his mother German. He studied mathematics and physics at the University of Vienna and attended the lectures of Hahn. Beginning his career as a teacher of mathematics he soon became an unofficial assistant to Moritz Schlick. It was Schlick's concern to see that the new philosophical ideas developed by Wittgenstein from the time of his return to philosophy in the later 1920s were made public that determined the subsequent shape of Waismann's activities. Until the outbreak of the war in 1939 his main task was the preparation of a book in which Wittgenstein's thought was to be systematically expounded. Between 1927 and 1935 this project was carried on in close personal conjunction with Wittgenstein. A first version of the planned book, *Logik, Sprache, Philosophie* seems to have been completed by 1931. A very different version came to England with Waismann in 1938. It finally appeared, in an English translation, as *Principles of Linguistic Philosophy,* in 1965, six years after Waismann's death. Had it been published when it was ready, fifteen years before the publication of Wittgenstein's *Philosophical Investigations,* it would have had the most notable impact. As it happened it attracted comparatively little notice. By the time it did come out the main content of Wittgenstein's later thought was already publicly available in Wittgenstein's own words and the publication of all Wittgenstein's unpublished material was already well advanced. Furthermore Wittgenstein's philosophy had to some extent lost its hitherto indisputably commanding position in the centre of interest of the English-speaking philosophical world.

Behind this set of objective historical facts lies a story of great pathos, even a tragedy, a philosophical analogue of Flaubert's *Un Coeur Simple*. Under the influence of his admiration for Wittgenstein and his personal devotion to Schlick, Waismann, at the crucially formative stage of his career, gave himself over wholly to the task of organising and expounding the ideas of someone else. By 1935 Wittgenstein was no longer willing to collaborate directly on the task. When Waismann came to England and sought to renew the collaboration Wittgenstein rebuffed him. The other source of Waismann's self-denying commitment, Schlick, had been murdered in 1936 and before his death his interests were already moving away from the area of Wittgenstein's main concern. Waismann, thus had to undergo a double intellectual and personal deprivation in addition to the ordinary spiritual dislocations of exile.

Waismann was not the only person to suffer from Wittgenstein's capriciously dictatorial temperament. But Waismann's achievements are such as to add a sense of loss as well as of sympathetic pain to the contemplation of his life. It seems clear that Wittgenstein benefited from all the work that Waismann did on his behalf. Waismann, despite a cultural predilection for the poetically indefinite and impressionistic, was a lucid and coherent expositor. In his drafts Wittgenstein would have found, not misinterpretation, which would have been understandable enough, given the profound originality and constitutional inchoateness of Wittgenstein's thinking, but as faithful a reflection of what it actually was at the time as could have been provided. If at first he could properly reprove Waismann for supposing that his thought in the late 1920s had not moved as far from the doctrines of the *Tractatus* as it had, in rejecting Waismann's work after that error had been corrected he was really expressing dissatisfaction with himself. It was as if Waismann was a kind of human notebook in which Wittgenstein could see, and respond critically to, the results of his own reflection.

The articles collected in this volume fall into two historical groups: the first, those written or published between 1928 and 1939, being with one exception the work of his time in Vienna, in close collaboration for a time with Wittgenstein and throughout with Schlick; the second composed entirely of unpublished work written in England in the 1950s. This feature of the selection gives a misleading impression of discontinuity. As his Aristotelian Society symposium paper of 1938, on the relevance of psychology to logic suggests, he was continuously productive from the moment of his ar-

rival in England. It is just the fact that his main writings in the period from
1945 to 1952 (the articles 'Verifiability' and 'Alternative Logics' of 1945, his
'Analytic/Synthetic' series in *Analysis* between 1949 and 1952 and his
paper 'Language Strata') are already easily accessible that excludes them.

Waismann's philosophical career falls into three main stages. The first
runs from 1927, when the frequent discussions with Wittgenstein, often in
the company of Schlick, began, until Wittgenstein broke off the collaboration in 1935. The chief published fruit of the work of this stage was Waismann's very Wittgensteinian *Einführung in das mathematische Denken* of
1936. During this period the first version of Waismann's general account of
Wittgenstein's philosophy, *Logik, Sprache, Philosophie* seems to have
been completed by 1931. During the previous year it had been announced
as the forthcoming first volume of the Vienna Circle series *Schriften zur
wissenschaftlichen Weltauffassung*. The advertisement described it as a critique of philosophy through logic. 'This work', it went on, 'is in essentials a
presentation of the thoughts of Wittgenstein. What is new in it and what
essentially distinguishes it is the logical ordering and connection of these
thoughts'. The contents were set out as follows: 1. Logic (meaning, reference, truth, truth-functions, essence of logic); 2. Language (analysis of
statements, atomic sentences, logical representation, limits of language); 3.
Philosophy (application of results to philosophical problems). No copy
seems to have survived.

In the period from the middle 1930s to his departure for England, the second stage, Waismann published a number of articles in *Erkenntnis* (those
on identity, logic as a deductive theory and logical analysis that are chapters III, VI and VII in this volume) but was mainly concerned with complet-
ing the revised version of *Logik, Sprache, Philosophie* in the form in which
it was eventually published.

The third stage runs from his settlement in Oxford around the beginning
of the war until his death there in 1960. Waismann was a respected but
somewhat isolated and eccentric figure in the Oxford philosophical com-
munity during those two decades. The most evident cause of this was linguistic. Waismann's command of English was far from secure, for all the effort he put into it. His syntax was often peculiar, as in the arresting first sentence of his well-known philosophical *Selbstdarstellung* in *Contemporary
British Philosophy, Third Series*, 'How I see Philosophy', which runs;

'What philosophy is?' Phonetically he was at times impenetrable and always memorably idiosyncratic. The word 'word' was always 'vürd', assertions of natural regularities were referred to as 'kozal loce'. Stunned by the complexity of English idiom he always carried a notebook with him in which to record new examples of its multifariousness. But his graduate classes were steadily attended and his papers to the philosophical society drew large audiences. Intense suspicion of English, and perhaps particularly college, food helped to diminish his involvement in the social life of the place.

Since Waismann's philosophical work was so dominated by the task of organising and expounding Wittgenstein's ideas there is a natural tendency to consider him as nothing much more than a convenient channel through which the thoughts Wittgenstein somehow found it impossible to communicate, except by word of mouth or in the privacy of unpublished writings, could reach a larger audience. Like others close to Wittgenstein, Waismann was left in no doubt as to the level of intensity with which indebtedness should be acknowledged. There was an angry response by Wittgenstein to what he regarded as the inadequate admission of dependence given, after their collaboration had been broken off, in Waismann's *Erkenntnis* article of 1936 on the concept of identity (Chapter III below). Waismann said, 'For valuable suggestions in developing the present view, the author is indebted to numerous conversations with Mr. Ludwig Wittgenstein - concerning, among other things, the concept of identity'. Not only is Wittgenstein thanked for suggestions rather than the whole content of his paper; there is also a certain irreverent playfulness, even flippancy, about that 'among other things'. At any rate Waismann took the rebuke to heart. Two years later, in 'The Relevance of Psychology to Logic', his tone is much more submissive: 'I wish to emphasize my indebtedness to Dr Wittgenstein, to whom I owe not only a great part of the views expressed in this paper but also my whole method of dealing with philosophical questions. Although I hope that the views expressed here are in agreement with those of Dr Wittgenstein, I do not wish to ascribe to him any responsibility for them'.

There are two main considerations which make Waismann more than a temporary, convenient channel of communication between Wittgenstein and the philosophical reading public. The first of them is that the order and coherence Waismann gave to the ideas he got from Wittgenstein add up to a major positive gain. The rambling and disconnected nature of Wittgenstein's own exposition makes Waismann's work much more than supererogatory. Secondly, there is the matter of Waismann's own development of ideas of ultimately Wittgensteinian inspiration. His doctrine of language strata, of conceptual domains within which strict logical relations obtain but between which and neighbouring domains only looser, 'criterial' relations prevail, is a valuably systematic extrapolation of Wittgenstein's comparatively indiscriminate conception of a language as a collection of innumerable, more or less autonomous language-games. His account of the 'open texture' of the expressions of ordinary discourse, the fact that the rules that govern their application cover only more or less familiar contingencies and do not provide for surprising, but envisageable, possibilities, is a valuably specific development of Wittgenstein's idea that language is not a calculus. Waismann's point is not so much that words of common speech are vague, that there are borderline cases in which we cannot decide whether to apply them or not, though he would not have denied that; it is rather that the operative criteria for their application are in practice only satisfied when certain other conditions, not included among those criteria, are satisfied as well. What we should say in a conceivable case where the criteria are satisfied but the ordinarily accompanying conditions are not is thus indeterminate. A direct consequence of this thesis is that it often cannot be decided in the light of existing rules whether a sentence is analytic or synthetic, the theme of his *Analysis* series of articles between 1949 and 1952. Is it an analytic truth that I see with my eyes? If an eyeless person could non-inferentially determine the colours of things by touching them with his fingers should we say he saw them or not? As it is, non-inferential colour-discrimination and the possession of a functioning optical apparatus seem always to occur together. But they are distinguishable and the first could occur without the second.

Another inference from the doctrine of open texture is that the kind of linguistic indeterminacy it implies is a positive advantage. It allows for the continuous development of a language to accommodate new discoveries, as exemplified by the progressive amplification of the scope of the concept of number from the positive integers to complex numbers. It might be concluded that this process of historic amplification is what lies behind the phenomenon, more stressed than explained by Wittgenstein, of 'family resemblance'. A final example of a substantial and influential development by Waismann of an idea suspended inexplicitly and, as it were, in solution, in Wittgenstein's thought, is his account of facts. The picture theory of the *Tractatus* had held the world to be composed of facts, conceived as fixed and objective articulations of reality. To reject that view, Waismann saw, involved a new notion of fact, one which he conveyed in the image of a cut made in the continuous fabric of reality by the linguistic and conceptual instruments men have chosen to employ.

These are the ideas through which Waismann's influence on philosophy has been principally exerted, by way of their publication in English in the 1940s and early 1950s. The present selection of hitherto less available or unavailable writings from the periods before and after that of his familiar English publications show the development of the doctrines in the earlier phase and the application of his methods to new topics in the later one.

The first chapter, on the axiom of reducibility, is uncharacteristically formal and briskly conclusive. Its aim is to show that the axiom of reducibility, which everyone, including its original propounders, has found to be far from intuitively obvious and selfevident, is not a tautology. Waismann attempts this by proving the consistency of four assumptions, the possibility of a world with (i) an infinity of individuals with (ii) an infinity of predicative properties each, in which (iii) no two individuals share all their predicative properties and (iv) no predicative property is possessed by only one individual. In any world, it might be said, however many individuals existed or properties were exemplified in it, the third and fourth assumptions taken together would imply that some non-predicative properties would not be coextensive with any predicative property. Assumption (iii) entails that the set of an individual's predicative properties is proprietary to it and thus that each individual has, and is the only individual that has, the non-predicative property of having just that proprietary set of predicative properties. Assumption (iv), that there are no proprietary predicative properties, entails that there are no predicative properties that are coextensive with the proprietary non-predicative properties entailed by assumption (iii). But are assumptions (iii) and (iv) generally consistent with each other? If *being F* and *being G* are predicative properties of an individual *a*, it surely follows that *being F and G* is also a predicative property of *a*. So, if the simple predicative properties of *a*, *being F, G, H... N* are finite in number then to the non-predicative property of having all *a*'s predicative properties there corresponds the predicative property of *being F and G and H and...and N*.

With the second chapter, on the concept of probability, the influence of Wittgenstein becomes apparent. It is most explicit and extensive with regard to the range theory of probability, ultimately derived from Bolzano, which Waismann carries on from the point it had reached in the *Tractatus*. But more noteworthy is the statement on the second page of the chapter: 'the sense of a proposition is the method of its verification'. This is the first published statement of that famous slogan and its source would appear to be Wittgenstein himself. Waismann makes it in the course of criticising the none too clearly defined theory that probability is an 'ultimate category of thought'. He goes on to reject the frequency theory. His first argument against it is unpersuasive: that it cannot explain why we do not expect the faces of a true die whose centre of gravity has been shifted to fall unequally on different faces. A supporter of frequency would reply that such an expectation would not be rational unless we had observed that the fall-frequencies of a die, hitherto in the habit of falling equally often on each of its sides, became unequal after its centre of gravity had been changed. More convincing are Waismann's objections to the identification of probability with the limit of relative frequency in an infinite series of cases.

Waismann goes on to propose instead that the probability one statement gives to another should be defined in terms of their comparative 'scopes', in terms, as he puts it, of the 'logical proximity' of the two statements. The probability p gives to q he defines as the scope-measure of p *and* q divided by the scope-measure of p. The assignment of scopemeasures is left somewhat indeterminate. It is treated as a matter of choice and the choice could, in appropriate cases, fall on frequencies, a suggestion hard to square with the notion that the probability-relation is a matter of purely logical proximity. The programme sketched in this essay was not carried out at all thoroughly until Carnap's *Logical Foundations of Probability*.

In the third chapter, on the concept of identity, the familiar phrase 'criterion of identity' makes its first appearance in the sense which it now commonly possesses. To ask an identity-question, he says, may be to call for a decision about a criterion of identity rather than to ask whether it has been satisfied. Many nowadays would reject his view that there is no one strict sense of identity, that 'the same' has as many senses as there are criteria of sameness. With the *Tractatus* and against Frege, Waismann holds that if '*a*' and '*b*' are signs for the same object then the statement '*a* = *b*' is a tautology.

The fourth chapter, on Schlick's significance for philosophy, presents in now familiar terms the distinguishing features of the work of the Vienna Circle as Schlick, rather than, say Carnap, saw it. Schlick, Waismann says, drew attention to philosophical questions, rather than to the answers proposed for them. Philosophical problems are signalled by a characteristic kind of perplexity; they are not solved by the acquisition of new knowledge but, arising as they do from lack of understanding more than ignorance, are dissolved by bringing to the surface the logical grammar that everyone knows but often fails to apply.

It is with the long and, until this volume, unpublished, paper on hypotheses that Waismann's detailed development of Wittgenstein's later thought effectively begins. A hypothesis is not just a description of fact, distinguished from others by its generality and thus its uncertifiable nature. Its sense is to be found in the work it does, but it is not just a recipe for prediction since it falls within the scope of logic. Particular observations are loosely related to hypotheses. We do not drop a hypothesis because of a single contrary observation; in general, observations tell for or against hypotheses, they do not prove or refute them. Thus an observation-statement does not follow from a hypothesis alone, but only from the hypothesis in conjunction with something else and that something else is not clearly specifiable, but is an amorphous background assumption of the absence of disturbing influences. It follows that 'verification' has different sense in application to hypotheses and to observation-statements. Waismann goes on to speak for the first time of different strata of language which interact logically with each other, even if not related by logically rigid connections. We fix the sense of our hypotheses by what we do with them in the face of apparent refutations. A hypothesis we preserve against all such refutations becomes thereby a 'form of representation', a conventionally analytic truth. Change of meaning and change of theory, it is implied, go ahead indiscriminably together.

In a way Waismann could hardly have chosen a less favourable example than that of the relation between a hypothesis and a singular observation to make his point about language strata, for he must have had in mind a simple hypothesis containing the same general terms as the observation-statement – no one would suppose that a single observation could by itself overthrow a hypothesis in which unobservable theoretical entities were mentioned. It is true that promising hypotheses are not usually abandoned in the face of a single counter-example. But that does not show a lack of strict logical connection between them. Rather the counter-example is rejected, explained away as some kind of mistake. Nevertheless the general idea he derived from this ill-chosen example proved capable of illuminating development.

The point that by changing the axioms of a logical calculus one simply gives new senses to the formal terms it contains is taken, in the sixth chapter on logic as a deductive theory, to show that there are really no alternative logics, a conception he was to consider more sympathetically a few years later in his Aristotelian Society paper on the subject.

The seventh chapter, on the relevance of psychology to logic, is part of a symposium discussion with Russell in 1938. Waismann expounds Wittgenstein's view about the incorrigibility of avowals of immediate experience, whose grammar is such that falsity and mistake are not allowed for, only slips. He concludes that such avowals are not really propositions and that, since the possibility of doubt is not provided for, they are not possible objects of belief. He criticises Russell's view that the logical word 'or' gets its sense from its connection with feelings of hesitation and doubt and also Russell's attempts to give a causal account of meaning, instead of interpreting it in terms of the following of rules.

A lucid and useful survey of the view prevailing in the late 1930s about the main assumptions of the *Tractatus* is supplied in the eigth chapter on logical analysis. He accepts the reductionist programme in its main outlines but reinterprets it in accordance with his conception of the loose, criterial, relations between different language-strata, of which that between discourse about perceptual experiences and material things is now put forward as a prime example. Against the original Wittgensteinian notion of elementary propositions, he claims that they need not be fully determinate and are not logically independent of one another.

Discussing fiction in chapter IX, he denies that the theory of descriptions applies to it on the ground that it is beyond truth and falsity, except in a marginal way where the fiction is part of a cultural inheritance, as when we might say that it is false that Hamlet was a married man. Names in fiction are dummies, used for pretended reference; they are not cases of failure of reference.

In the following note on existence he discusses Russell's claim that 'this exists' is meaningless. He sees it, rather, as a kind of tautology and rejects Moore's argument that since one can significantly and truly say 'this might not have existed' it follows that it is possible, even if false, that this does not exist. The statement that this might not have existed does not allow for the possibility that this, as pointed to now by the word 'this', does not exist.

It is with the eleventh chapter, on experience, that the final and most distinctive phase of Waismann's thinking, fully detached from Wittgenstein, makes it appearance. He begins, straightforwardly enough, by pointing out that the word 'experience' is not an ordinary classifying or distinguishing term since it has the paradoxial property of applying to everything. He then embarks on a brief cultural history of subjectivism, tracing it from the Reformation and the scepticism of Montaigne, through the invention of the concepts of the interesting and the boring in the eighteenth century, to its full theoretical flowering in the idealism of the nineteenth century. The notebooks in which his linguistic botanising was recorded come into play with a long list of the first occurrences, as registered by the O.E.D., of various words for characterising things by their relation to emotions: 'frightening', 'exciting' and so on. The moral he draws from this historical excursus is that the main role of the philosopher is to crystallise, in an ostensibly theoretical form, the idea-movements of his age. He does not recoil from going on to draw the kind of logically antinomian conclusions to be found in 'How I See Philosophy'. Philosophy is essentially a matter of vision. The philosopher should not let himself be disturbed by the question of whether what he says is true or false. 'To ask whether some metaphysical vision of the world is right or wrong is almost like asking whether, e.g., Gothic art is true or false'. With the bit between his teeth he goes on to say that insistence on logical flawlessness and 'the belief that you can "refute" a philosophy by digging out some internal inconsistencies' is a relic of scholasticism. 'A philosopher', he concludes, 'may write a book every sentence of which is, literally, nonsense, and which none the less may lead up to a new or a great vision'.

Waismann's discussion of the linguistic technique in chapter XII is a critique of the then current preoccupation of philosophers with the conventional rules of ordinary speech. To study the actual use of language is not enough. The philosophic analyst must consider the phenomenon itself as

well as the expressions that are applied to it and he must seek out the underlying rationale of the rules he elicits.

The last two chapters are somewhat fragmentary discussions of various
issues arising about knowledge and belief. In the first of them he argues that
the two are different in kind, not merely degree, and leads up to that conclusion with an album of varieties of belief: those that are positively adopted
and those that simply emerge; those that are connected to action and those
that are not. In the second he reconsiders the main conclusion of the earlier
piece and admits that there are cases where knowledge is no more than a limiting case of belief, as when it comes to be claimed as the result of a steady,
continuous accumulation of favourable evidence, as well as cases where it is
not, as, for example, where a new attitude comes to be adopted, a stubborn
unwillingness to hear further argument on the topic in question.

Personal tragedy, the unhappy deaths of his wife and son, darkened
Waismann's last years and must have augmented his sense of isolation. The
distinctive antirationalist tendency of his last thoughts was never more
than schematically intimated. It was welcomed by many who chafed at
what they saw as a kind of lexicographic despotism but it left many questions open. Was the more or less late-Wittgensteinian kind of philosophy
that he himself had been practising no more than a crystallisation of the
idea-movements of his age? At any rate his view reinterpreted the history of
philosophy not as a sort of conceptual disaster-area to be picked over in the
diagnostic style of an accident investigator. Most promising for future development was his proposal that philosophers should seek the underlying
rationale of the linguistic practices they examine, a proposal followed in
the concern with transcendental arguments that has been characteristic of
the period since his death. Even for those who can see in him no more than
an expositor of Wittgenstein, he has something of value to offer, a commitment to clarity and order, a readiness to take a definite stand, whose absence from the latter writings of his master often makes them so exaspera-
tingly difficult to interpret.

ANTHONY QUINTON

EDITOR'S NOTE

This volume brings together all of Friedrich Waismann's completed papers that have not previously appeared in book form in English. Of particular importance for the present series are his five pre-war articles published in German and now for the first time translated (on the axiom of reducibility, on probability, on identity, on logic as a deductive theory, and on logical analysis; forming Chapters I, II, III, VI, and VIII). These were valuable contributions to the work of the Vienna Circle and well illustrate that the so-called Circle was more of an ellipse and had Schlick and Waismann as one of its epicentres, Carnap and Neurath, perhaps, as the other.

Three other pre-war papers are a tribute to Schlick (first published in Dutch), a chapter on hypotheses from Waismann's book intended to expound Wittgenstein's philosophy (a chapter not published even in German before the present year), and the paper on psychology and logic that Waismann gave to the Joint Session of the Aristotelian Society and the Mind Association in 1938 (practically his first appearance on the English philosophical scene, in which he was to spend the rest of his life). These form Chapters IV, V, and VII. The last two lean heavily on the work of Wittgenstein. One of them went unpublished because, no doubt, the positivist phase of Wittgenstein's thought seemed to be superseded. But interest in it has now reawakened.

The remaining six chapters (on fiction, on existence, on experience, on linguistic philosophy, on belief, and on knowledge) are post-war papers of Waismann's, composed in English and for the most part delivered to philosophical societies and discussion groups in Oxford. None has previously been published. There is little overt controversy, but Waismann's reactions to the publication of Wittgenstein's *Philosophical Investigations,* to the teachings of J. L. Austin, and to the early articles of P. F. Strawson are easily seen. The characteristic topics of the 50's are discussed with exemplary clarity, with due scepticism, and with that exceptional perspective which Waismann's long involvement with analytic philosophy had given him.

Apart from indications of provenance footnotes have been relegated to the end of a chapter, in accordance with the publisher's practice. Those supplied by the present editor are enclosed in square brackets. They will be found to be few. No general attempt has been made to supply precise references where Waismann himself gave none. Often this would have involved guesswork; yet more often it is clear that onyly a general reference was intended.

The publication of the present volume was made possible by permission from Waismann's literary executors, Sir Isaiah Berlin, Mr Stuart Hampshire, and Professor Gilbert Ryle. Deep thanks for the preservation and publication of Waismann's papers are owed to them all, but particularly to Professor Ryle, who acted for the three. He died while this volume was in the press. May it contribute to his memorial.

B. McGUINNESS

CHAPTER I

THE NATURE OF THE AXIOM OF REDUCIBILITY*

The so-called axiom of reducibility plays a prominent part in the reconstruction of mathematics, as carried out in *Principia Mathematica,*[1] the basic work of the logical school. The axiom says: *for any propositional function* $\varphi\hat{x}$ *there is always a predicative propositional function* $\psi!\hat{x}$ *which is formally equivalent to* $\varphi\hat{x}$.[2] The role of the axiom of reducibility is to make possible the transition from any function to a predicative function and thereby to remove the difficulties entailed by the setting-up of the so-called ramified hierarchy.[3] Within the framework of *Principia Mathematica,* such an axiom is indispensable. Without it or some equivalent axiom, the theory of real numbers, as developed there, would collapse, and with it the foundations of analysis. On the other hand, in setting up the axiom, Russell and Whitehead had felt instinctively that it had a different character from the other axioms of logic and that the same degree of certainty could not be ascribed to it; but since the real nature of the axiom remained obscure, the authors were content to arrange their work in such a way that it was possible to recognize at each point which propositions depended on the axiom of reducibility and which not.

The propositions of logic have a characteristic property which distinguishes them from all other propositions: they are tautologies. A tautology is a proposition which is true merely on the basis of its structure. To be more precise: a tautology is a truth-function which remains true under all truth-value distributions of its arguments.[4] A proposition of logic, therefore, does not communicate a state of affairs. To communicate a state of affairs, we can only use a proposition which has the capacity for being false (when the state of affairs asserted in the proposition does not exist). This is also the reason why the propositions of logic are exempt from testing through experience. For testing a proposition through experience means comparing it with reality (with a real state of affairs). A tautology, how-

* First published in German in *Monatshefte für Mathematik und Physik* **35** (1928), 143–6. Reprinted in Reitzig (see [31] in Bibliography on p. 188).

ever, expresses neither the existence nor the non-existence of a state of affairs; it cannot therefore be either confirmed or refuted by experience. To speak with Leibniz, a tautology must be true in any world, not just in *ours*. If a proposition is true only in our world, this is a sure sign that it is not a tautology and that it does not belong to logic.

The difficulty felt by Russell and Whitehead can now be expressed precisely in this proposition: *The axiom of reducibility* – unlike the axioms of logic – *does not represent a tautology*, and there is no excuse for introducing it into logic. The aim of the following discussion is to demonstrate this fact.

If the axiom of reducibility is a tautology, then it must hold for any world, not just for ours. Thus if we can construct a world for which the axiom of reducibility does not hold, we shall have shown thereby that it is certainly not a tautology. Let us imagine a world with the following properties:[5]

(1) There are infinitely many individuals in it.

(2) Each individual has infinitely many predicative properties.

(3) No two individuals have all their predicative properties in common.

(4) Whenever a predicative property belongs to one individual, it also belongs to some other individual. In other words, there is no predicative property belonging exclusively to one individual.

If 'a' is the name of an individual, then the class c of individuals who have all their predicative properties in common with a can be defined by

$$(\varphi) \, . \ \varphi!\hat{x} \, . = \, . \ \varphi!a.$$

The propositional function we have constructed is certainly not predicative (for it appeals to a totality of predicative properties); let us refer to it as $\varphi_1\hat{x}$. According to the axiom of reducibility, there would then have to be a predicative propositional function $\psi!\hat{x}$ which was formally equivalent to $\varphi_1\hat{x}$:

$$\vdash : (\exists\psi) : \ \varphi_1 x \, . = \, _x. \ \psi!x$$

But there can be no such predicative propositional function. For according to assumption (3), no individual agrees with a in all its predicative properties; class c therefore contains only the single individual a. On the other hand, if a predicative property belongs to a, it also, according to requirement (4), belongs to some other individual. It follows that there is no predicative function coextensive with $\varphi_1\hat{x}$ – so the axiom of reducibility has failed.

The only question that remains to be examined is whether the world we have just constructed may not be contradictory, i.e., whether our four requirements are compatible. For this purpose, let us project the objects of our world into another realm which has already been shown to be free from contradiction; our four requirements cannot, then, lead to a contradiction either.[6] Let the realm R into which we project the objects of our world be the realm of rational numbers. Let a rational number correspond to each individual. Let classes of rational numbers correspond to the predicative properties of individuals according to the following prescription: to the rational number r shall correspond all open intervals containing r and bounded by rational numbers, in so far as these intervals fall into realm R, i.e., the intersections of these intervals with R. Requirements (1) to (4) are then satisfied. Requirement (1) says: there are infinitely many rational numbers; requirement (2): each rational number r lies in infinitely many intervals; requirement (3): no two rational numbers lie together in all intervals; requirement (4): each open interval contains more than one rational number. This shows our requirements to be free from contradiction.

In conclusion, it should be emphasized that we have confined ourselves deliberately to rational numbers and constructible classes, in order to avoid the difficulties which lie in the concept of a real continuum and the concept of an arbitrary subclass respectively – difficulties which are partly due to the character of the axiom of reducibility.

NOTES

[1] Whitehead and Russell, *Principia Mathematica*, Cambridge, I, 1st ed., 1910, 2nd ed., 1925, II, 1912, III, 1913.
[2] *Loc. cit.*, 1st ed., pp. 59 and 174.
[3] *Loc. cit.*, 1st ed., p. 51 Cf. also Ramsey, *Proc. London Math. Soc.*, II, 23, 1925, pp. 356 and 358 ff.
[4] Cf. Ludwig Wittgenstein, *Tractatus Logico-Philosophicus*, 4.46.
[5] Cf. also Ramsey, *loc. cit.*, p. 381.
[6] I am indebted to Mr. R. Carnap for suggesting the following proof.

CHAPTER II

A LOGICAL ANALYSIS OF THE CONCEPT OF PROBABILITY*

The aim of the following dicussion is the logical clarification of the concept of probability. I want to give a definite answer to the question: What does probability mean, and what is the sense of probability statements? In accord with Leibniz and Bolzano, I believe that the theory of probability is a branch of logic. And using Wittgenstein's ideas, I should here like to explain how this view can be freed from the difficulties which have till now stood in the way of its acceptance.[1]

The word probability has two different meanings. Either we speak of the probability of an event: this is the sense in which the word is employed in the probability calculus. Or we speak of the probability of an hypothesis or law of nature: in the latter sense, probability is only another word for the utility of an hypothesis or law of nature, i.e., for the inconvenience it would cause us to replace the hypothesis by another. The two meanings have nothing to do with each other, and it would be better if we employed two different words in speaking of them. I shall speak here of probability only in the sense of the probability calculus.

There are today two general positions one can adopt on the foundations of the probability calculus. One view, widely held in philosophical circles, is that probability is an ultimate category of our thought. To explain this view more fully, it is held that we can never be sure whether certain statements are true and that we can make those statements only with probability. If we carry out a series of measurements to determine, e.g., the length of a stick, the results we get will, as is well known, all differ a little from one another. It is now said: we can never decide what the 'true length' of the stick is; any given length has only a certain probability, and it is at this point that the concept of probability enters into physics. This idea has been pursued so far that the foundations of logic have been abandoned just to make room for the concept of probability; thus it has been alleged: it is mere prejudice to

---

* First published in German in *Erkenntnis* 1 (1930–1), 228–48. Reprinted in Reitzig.

assume that a statement can only be true or false; a statement can also be probable. But in my opinion this is merely playing with words. A statement describes a state of affairs. The state of affairs exists or it does not exist. There is no third thing, and hence also no intermediary between true and false. If there is no way of telling when a proposition is true, then the proposition has no sense whatever; for the sense of a proposition is the method of its verification. In fact, whoever utters a proposition must know under what conditions he will call the proposition true or false; if he cannot tell this, then he also does not know what he has said. A statement which cannot be conclusively verified is not verifiable at all; it just lacks all sense – but this has surely nothing to do with probability. This applies also to our example, and it is not difficult to point to the source of the misunderstanding. It is in the little phrase 'the true length'. What we have actually observed is the individual results of measurement. By using an arithmetical rule, we produce a new number – the so-called mean value – out of the given numbers and *call* it the 'true length' of the stick. The question whether the stick *really* has this length or not has no sense whatever since it treats a mere convention as if it were a question of fact. Anyone who has grasped the true origin of this concept and seen clearly that length is not something immediately given but derived in a certain way from observed data – in the same way, incidentally, as the other concepts of physics: the concept of fieldstrength, the concept of potential, etc. – will no longer be tempted to think that the stick must have a definite length 'in itself' and that we are merely unable to determine the exact value of this length.

The second view, the statistical one, equates the concept of probability with the concept of relative frequency. When I say, 'The probability of throwing 2 with a die is 1/6', I mean this: if one goes on throwing long enough, the number 2 will occur on the average every sixth time in the series of results. From the outset this theory avoids the rocks on which earlier theories were wrecked. The following problem, which earlier thinkers found so troublesome, simply does not arise: How can probability determine the actual course of events? Can pure mathematics, then, anticipate reality? We should not be surprised at this, for probability is defined precisely as relative frequency. The starting-point of this theory is in fact so well chosen that the theory can give an account of all observations and all actual applications of the probability calculus to statistics. But before we regard it as firmly established on this account, we must ask whether it also satisfies all legitimate demands. And then we notice how easy it is to raise questions which are definitely embarrassing to the theory. For example, we see a die lying on the table; we perceive that it is a 'true' die, that the probability of throwing 6 is 1/6; we now shift its centre of gravity towards one side; we expect that this displacement will also come out in the results, i.e., that the die will turn up more frequently on the one side. We are not deceived in our expectation; the die falls in fact as we had expected. But what right did we have to assume this? If we turn to the statistical theory for enlightenment, we are told: it makes no sense to ask for the reason for the occurrence of a particular relative frequency; statistical series with their relative frequencies are the ultimate facts which serve the probability calculus as its point of departure, and it must decline to go back beyond its point of departure; as is well known, there are no considerations leading beyond the statements of deterministic physics to the facts of statistics; so as far as one can see, the only task of the probability calculus is to derive further distributions from given initial distributions. Who will be convinced by such an argument? Who can be made to believe that there is really no connection here? If we had to take our stand on this issue, we should not be judging incorrectly if we saw in this negative attitude a confession of the weakness rather than the strength of the view. In fact, no question capable of being formulated clearly can be got rid of by being called unscientific. It may be convenient to postpone an unsolved problem for a while, but in the long run it cannot be rejected. We all feel that there is some kind of connection here. And we rightly tell ourselves that if that question admits of no answer within the framework of the statistical theory, then it is not because our question was unjustified but because one element is still missing in the structure of the theory and because the theory fails to grasp one of the essential features of probability.

The first serious objection to be raised against the statistical theory is that it has those consequences. The second is purely logical in nature and concerns the way in which relative frequency is formulated mathematically. Statistical observation presents us only with series of a certain finite extension. In order to arrive at its concept of probability, the statistical theory goes beyond the facts of the case in a certain way and imagines, e.g., the series of results of throwing a die to be infinite, i.e., it applies to this series the mathematical concept of a number series. But one thing is clear: what is infinite is only the law that generates the series. Every mathematician is perfectly familiar with the idea that to operate with an infinite series is at bottom only to operate with the laws that generate the series. When we speak, e.g., of the convergence of a series, we refer back only to that law and not to the 'actual' order of its members, of which we can never survey more than a finite number; and the same holds true for all other cases. In short, a mathematical series is essentially something lawful whose properties are perfectly surveyable. In comparison, nothing is clearer about a statistical series than its irregularity, and this alone proves that it is not a mathematical concept. In fact, anyone who stipulates that a convergent series of numbers be constructed in an irregular way – and this is what is at issue – stipulates the impossible; neither the mathematical series nor the empirical one can fulfil these requirements. Against this it must be said emphatically that a statistical series does not have the properties of a mathematical one; that the mark of the accidental, unforeseeable, cannot be transfered from the empirical structure to the mathematical one without destroying the *a priori* necessity which is the characteristic mark of mathematics.

The consequences of this logical muddle come out clearly in the formulations of the theory. When von Mises formulates the irregularity by saying that any arbitrarily chosen part of a series approaches the same limit as the whole, then this statement, taken literally, is certainly false. For among the partial series there are surely some that approach a different limit or do not converge at all. If it is said that it is futile in practice to make such a choice, that no player has till now succeeded in discovering a system for winning at Monte Carlo, then this is perfectly correct; but it only proves that we have left behing the realm of purely mathematical concepts.

The only way out of all these difficulties is to see clearly that irregularity cannot be formulated, that it makes no sense to first posit a mathematical series and then explain that this mathematical series obeys no formation rule, and that the non-existence of lawfulness would be expressed in a perfectly clear language by the non-formulation of laws and not through the laying down of an axiom of irregularity.

An adherent of the statistical theory will not, however, be convinced by these considerations. He will counter that it must make sense to admit such infinite series. Otherwise, how would it be possible to operate in physics with the laws of probability, which call for the constant use of the mathematical concept of a limit? The formulation of the theory may not have reached the final degree of clarity yet, it may still have to be supplemented and improved; but to doubt that such a view is ultimately tenable is impossible. It is true that we can try to operate with finite series; but in that case we shall never arrive at simple formulations. It is only the desire for simplicity and perspicuity which forces us to introduce the idea of an infinite series. And in doing this, are we not doing essentially the same thing as the other sciences, e.g. geometry, when it replaces the empirical forms given to the senses, the plane strips of finite depth with their imperspicuous relationships, etc., by perfectly exact forms on which it can operate according to its simple laws? The right enjoyed by geometry must also be granted to the probability calculus; if it is to arrive at a mathematically useful formulation, it too must idealize the given relationships, and for this purpose it needs the definition of a limit. This line of argument sounds convincing, and we would have to acknowledge its force if the analogy it adduces did in fact obtain. But is it correct?[2] That depends essentially on the concept of idealization. The usual view is this: If we take, e.g., a number of circles and measure the ratio of circumference to diameter, we get numerical values which, while imprecise, become more precise as we employ finer means of observation. It is then said: The number π is the ideal limit towards which the results of measurement tend as the precision of the measurements increases without limit. But is this view correct? Our misgivings begin with the word 'precise'. When do we call a measurement precise? When, e.g., the measurement of a circle, performed with extreme care, yields a different value for this ratio, we do not say: the propositions of geometry are false, experience teaches that the number π does not have the same value for all circles; what we say is: we have made a mistake, our ruler has changed shape, the figure was not a circle, and similar things. That is: whatever the measurement may turn out to be, we hold on to the number π and regard it as a rule for deciding when the measurement of a circle is to be taken as correct. The number π is our yardstick for measuring the goodness of observations. Thus the exact opposite is true: the number π is not the ideal limit which we get to know with more and more precision through actual measurements, it is the yardstick given in advance by which we judge the degrees of precision of a measurement. These considerations help to bring out a fundamental truth which might be put like this: The propositions of geometry are a system of rules which we constantly employ in actual measurements, and by which we decide, e.g., whether a given line is a straight line, a given body a sphere, etc. To put it briefly: These rules are the syntax of the concepts we employ in describing actual spatial relationships. One cannot approach the rules of syntax through actual observation. Idealization does not mean refining the actual observations in thought beyond limit. What idealization means is describing the observations in terms of concepts with an antecedently given syntax.[3] One does not arrive at the ideal, one starts out from it. To operate with mathematical concepts involving infinite processes, with limits, differential quotients, etc., makes sense only within a system of stipulations which we bring to reality. Where such a system does not exist, talk of 'idealization' has lost all sense.

The error of this view has now been exposed. It starts out with a completely false idea of idealization and imagines therefore that what applies to geometry can be carried over to statistics. There can be no question of this in reality: an empirical series is not a mathematical one, a relative frequency is not a limit, and idealization cannot perform the miracle of turning the one thing into the other. So the argument which may have seemed at first glance to have some validity becomes in reality an argument against that view, and it proves conclusively the untenability of the assumptions underlying those considerations.

The consequences of this error are very far-reaching. Since probability cannot be defined as the limit of relative frequency, the rules of mathematics cannot be applied to those alleged limits, and the whole calculus which has been raised on this foundation collapses. But without this calculus, contemporary physics would be wholly unthinkable. The way out is clear. It has been inferred that since physics needs the probability calculus in the form of certain analytic propositions, and since probability is defined as relative frequency, the concept of a series must be idealized. We on the other hand infer that since the concept of a series cannot be idealized, it also cannot be used for defining probability.

For these two reasons, the statistical theory cannot be regarded as a finally adequate solution to the problem of probability.

I would now like to develop the outlines of a third view which I believe to be free from the objections which can be raised against the previous ones. It differs essentially from those theories by focusing on a certain logical relation between statements, a relation which could be called the degree of 'logical proximity' of two statements. I shall begin by showing how one can arrive at the concept of probability through purely logical considerations by starting from the theory of deduction.

We start therefore with the remark that all the statements we make in everyday life or in the sciences are more or less indefinite. When I say, 'My friend is in Paris' or 'There is a red ball in this urn', the actual facts of the case may vary considerably without the proposition's ceasing to be true. We here encounter a peculiarity of our statements which can be formulated as follows: A statement does not simply specify a fact but a scope or range of facts. As long as the actual states of affairs move within the scope delimited by the proposition, the proposition remains true; as soon as they go beyond it, the proposition becomes false. The narrower the scope, the more exactly the proposition corresponds to reality and the more definite its sense. If I attribute a special significance to this conception of scope, it is because it is very important for logic and, in particular, for examining the question when one proposition follows from another. The answer to this question can be stated very simply by employing our mode of expression: one proposition follows from another when the scope of the one includes that of the other. Hence, in concrete terms, inference consists in narrowing the scope. The inferred proposition says less than the original one. When one proposition follows from another, the sense of the former is already contained in the sense of the latter.[4]

Let us call a relation of the kind portrayed an 'internal' relation. The word 'internal' is here meant to indicate that the relation is, as it were, already contained in the propositions and not just added on to them afterwards.

This relation is now capable of being generalized. When one scope includes another, we are dealing, as already indicated, with the relation of entailment. When the scopes wholly exclude one another, this means that the statements contain a contradiction. The general case is that one scope partially overlaps with another. It is this general case which now arouses our interest.

Entailment and contradiction are represented in this picture as – as it were – topological relations between scopes. Let us now take a further step and introduce a measure for the magnitude of a scope. We demand of such a measure that it have the following characteristics:

(1) if p is a statement and $\mu\,(p)$ its measure, then $\mu\,(p)$ shall be a real, never a negative number;

(2) an empty statement (a contradiction) shall have the measure 0;

(3) if p and q are two statements which are incompatible, then $\mu (p \vee q)$ shall equal $\mu (p) + \mu (q)$.[5]

These requirements do not yet determine the measure unambiguously; different systems of measurement are possible, and we can only choose between them from the point of view of convenience. Let us call those statements to which we have assigned a measure on the basis of a stipulation 'measurable statements'. These measurable statements can, among other things, be said to form a body.

We now come to the most important concept in our exposition, namely the concept of probability. If p and q are two measurable statements, then we shall understand by the measure of the probability which the statement p gives to the statement q the following quotient:

$$_p\text{prob}_q = \frac{\mu (p \cdot q)}{\mu (p)}$$

In concrete terms, this means that the probability is the magnitude of the common scope of p and q in proportion to the magnitude of the scope of q. The probability, so defined, is then, as it were, a measure of the logical proximity or deductive connection of the two propositions. When p entails q, the degree is 1; when p contradicts q, it is 0.

If we replace p by a set of statements, say $p_1, p_2, \ldots p_n$, we can ask for the probability which all these statements combined give to the proposition q. This probability is found by combining $p_1, p_2, \ldots p$ into a logical product P and by determining, exactly as before, the probability which this product gives to the proposition q. (It may be that the probability remains unchanged when further propositions are added.) Let P now be the product of all the propositions we know to be true (including the laws of nature we assume hypothetically):

$$_P\text{prob}_q$$

will then represent the probability which the whole of our present knowledge gives to the statement q. This is what is commonly called the probability of the proposition q.

Three different cases ought to be distinguished here. Either P is so constituted that we can derive q from it in a purely logical way: this is the case when we predict an event from known data on the basis of certain laws of nature. Or P contradicts q: we then say that such a process is excluded by the laws of nature. In neither of the two cases do we commonly speak of probability. Probability begins only where we have some knowledge of the conditions under which an event occurs but where this knowledge is insufficient for making definite statements. Giving the probability means that some part of the presuppositions under which the event occurs is satisfied. The event itself occurs or it does not occur, and the statement that describes the event can only be true or false accordingly. 'Probable' is not an intermediary between true and false. What is probable is not the proposition, but our knowledge of the truth of a proposition. This knowledge is capable of different gradations, which is precisely what our formula expresses.

It is clear that this view has nothing to do with subjectivity; for what it brings out is the logical relations between propositions, and no one will want to call these subjective. If it is thought that the subjective element consists in this: that probability is made to depend on the state of our knowledge, then we reply: There is no other occasion for introducing the concept of probability besides the incompleteness of our knowledge. Everybody knows that the theory of probability is brought in only where our knowledge of events comes up against a limit; where we have some general knowledge of the conditions under which an event occurs but lack a detailed knowledge of what is involved. Determinism, the idea of a closed system of causal laws, remained an ideal in physics only as long as it was possible to think that one could advance to a knowledge of all the details; and if we now witness with increasing frequency the appearance of lines of thought belonging to the statistical mode, it is because it has been realized in the meantime that this ideal is in every respect unattainable. So the fact remains that probability begins where we have been denied certainty. But in that case, a formula which gives prominence to this state of affairs cannot be dismissed as subjective. This objection would be justified only in the following case: if it should turn out that the measure of probability were variable; that a probability we had calculated at one time could be overturned again as new facts came to be known, and without our ever coming to a definite conclusion. It will appear from the following that this case does not arise.

I believe that this view is a satisfactory foundation on which to construct the theory of probability. I shall first examine the logical justification for this approach; I shall then go on to its applications and show that our definition really captures what is meant by the word probability; finally, I shall discuss the relation of this view to the statistical theory and try to demonstrate that it is free from the difficulties which are attached to the latter and which make it an unsuitable foundation for the probability calculus.

To demonstrate that this approach is logically justified, we prove that all the propositions of the probability calculus can be derived from our definition by purely logical means and without further premises. From our definition follow all the axioms which served Keynes, for example, as the basis for his formalization of the theory of probability,[6] and from these in turn follow all the other propositions. The whole formal structure of the probability calculus is in fact contained in our definition.[7] But this formal structure appears to us in a somewhat different light. On the usual view, it looks as if the axioms of the probability calculus described something. But this inevitably raises the question what their truth is based on: whether on their inherent self-evidence or on experience or on something else. If we are not mistaken, the unclarity resides in the chosen manner of expression. One just cannot introduce axioms without giving rise to the idea that these axioms communicate something; that they are true because the state of affairs they describe exists. Our view keeps such interpretations at a distance. In deriving all the axioms from the chosen definition, it avoids making it appear as if the axioms stated something. This also throws some light on the question whether the probability calculus can be changed. Only a proposition with sense can be changed, for changing a proposition consists precisely in changing its sense. The theorems of the probability calculus refer to nothing and describe nothing: they are purely logical in origin. While there are different systems of mechanics and geometry, there is only one arithmetic, only one logic, and only one probability calculus. Although this statement ought to be obvious, it has been obscured by the view that the probability calculus is an empirical theory; that it describes statistical phenomena just as mechanics describes the processes of motion. Why exactly is this theory called empirical? If what is meant is that the relative frequencies are determined in an empirical way, then it is obviously true; in mechanics, too, the initial states, the masses, velocities, and distances are determined empirically. But these determinations do not yet amount to mechanics. The essential point is that mechanics derives further data from these data, that it infers future observations from present ones. If it is now thought that the same holds true for the probability calculus, that it derives further distributions from given distributions, then it must be said that the new distributions do not go beyond the old ones in content, that they add nothing to our knowledge, and that they are not prophecies. The propositions of mechanics state what we are to expect, and we can find out by experiment whether our predictions are true; but the probability calculus is only a method for expressing in a different form what we already know. What is verified by statistical observation is always our initial assumptions, never the probability calculus itself. When a physicist deduces a consequence from his theory and confirms it by experiment, no one will want to assert that he has thereby verified the validity of logic. But in the case of the probability calculus it is thought even today that its correctness depends on experience and that it could be changed through experience. It should count as obvious that just as a physicist does not check the validity of logic with his experiments, so nothing can be setteld about the theorem of addition through statistical experiences.

I should like to touch here on only one more point. It concerns the fact that there exists some freedom in our definition regarding the choice of a system of measurement. We must therefore be prepared in advance for the fact that the probability will turn out differently according to the assumptions we have made. As is well known, this actually happens in the case of Bertrand's paradox. (In its best-known version, this is the question: How great is the probability that a chord drawn arbitrarily in a circle will have a length greater than r? The answers run, according to the assumptions one has made: $2/3$, $3/4$, $\sqrt{3}/2$.) The crux of the matter is that a set of chords cannot be measured without further presuppositions; the different solutions of the problem correspond to so many specializations of the metric. In representing this result as the natural consequence of our definition, even as something that could have been expected all along, our view divests this fact of the air of paradox which has clung to it till now.

Thanks to this definition, we now have a clearly circumscribed concept of probability. We have also indicated that it leads to the desired formulas of the probability calculus. But, it will be asked, is this the concept of probability we use in applying the probability calculus? This question will now be discussed.

All applications of the probability calculus presuppose that an event can recur any number of times. A series of events gives rise to the concept of relative frequency, and our question is this: How is the concept of probability related to the concept of relative frequency?

We meet here with two very different cases, and our answer will accordingly be different. We may not understand the mechanism responsible for the occurence of the event. In such a case we have to rely only on statistical observations. We know, e.g., that 52% of all births are male; beyond this we know nothing. The statistical statement will then appear as the antecedent P of our formula, which gives the event – a male birth – the probability 0.52. In this case the probability reduces simply to the relative frequency, and our approach coincides with the customary statistical treatment of this problem.[8]

The situation is very different in what used to be called cases of a priori probability: games of chance may serve as representative examples. Here it is a matter of determining a definite probability from data about the conditions of the game and the known laws of physics. Let us consider, e.g., the game of roulette, where a ball rolls in a circle which is divided into narrow sectors which are painted alternately red and black. The scopes of the statements are here evidently the areas in which the ball will rest after it has come to a stop. If we regard equally large areas as equiprobable, we can calculate the probability that the ball will come to rest in, e.g., a red sector. Against this it will surely be objected that we do not know in advance which areas are equiprobable, and that this is where the whole problem of the application of the probability calculus lies. But this objection does not go as deep as it may seem. For it is possible to give the principle that has been applied a perfectly correct formulation. It should strike us that in all cases where such an a priori probability has been calculated the occurrence of the event depends on *spatial* relationships. These spatial relationships are capable of being measured. We have pointed out that our requirements do not yet determine the measure of a scope unambiguously; that we can choose the metric. It goes without saying that if the field is wholly unknown, we also do not know what metric applies to it; we must first stipulate one, and we stipulate a metric in such a way that our statistical experiences will agree with it. Logical considerations alone will not, of course, give us any knowledge of the behaviour of real things. We need not therefore fear the charge of apriorism. In fact, it is only from an immense sum of experiences that we know that the course of the ball does not depend, e.g., on the directions of the compass, the colour of the area, or similar things, and this is why we regard ourselves as justified in taking equally large areas to be equiprobable. Nor are we open to the other charge: that our view envisages a mere convention which must must be laid down anew in each case and which would then, of course, be of no value. The fact of the matter is this: Once a metric has been laid down for a particular field of phenomena, it then governs that field; we can then calculate the probability for any situation without first consulting our statistical experience, by adapting our formula to the specific conditions of the task; in a word: we can anticipate statistical experiences. Here lies the difference between our view and the statistical one. In cases where we are entirely dependent on statistics, our approach coincides with the statistical point of view. But it goes beyond it as soon as the circumstances allow us to set up a metric for the magnitude of scopes and to determine the probability on the basis of this metric.

Let me try to explain this by an example. Imagine a slightly inclined board with pegs attached to it at small regular intervals; we let little balls roll down the board between the pegs; in rolling down, the balls hit the pegs, change their course each time, and finally come to rest at the bottom, where they form the familiar bell curve. The probability calculus assumes the exact form of their distribution is described by Gauss's law. How is this statement to be interpreted? The statistical interpretation is this: Gauss's curve represents an idealization of the actually observed distributions, i.e., it is the limiting position which the actual arrangements approach without limit; if we could only make the number of pegs as great as we liked and repeat the trials as often as we liked, we would find this law confirmed with more and more precision. Our view on the other hand is this: While experience teaches that the actual distribution is approximately Gauss's, it does not, nor can it teach, that Gauss's is the limit towards which the actual distributions tend. The true state of affairs is this: When the ball hits a peg, it can deviate either to the left or to the right. What this depends on escapes our knowledge. We now stipulate: the two possibilities shall be equiprobable. In so doing, we stipulate a definite metric, and this metric – and not the actually observed frequencies – is the starting-point for all further mathematical considerations. These teach us to calculate the probable distribution for any number of such pegs. The result of this calculation is a definite formula, which contains the number of the series of pegs. With this formula – and not with the observations themselves – we can proceed to the limit by letting that number increase to infinity. We then get Gauss's law.

This view purifies the mathematical formulation by removing the unclarities which were attached to it earlier, and it does away with the opinion that, strictly speaking, only an infinite number of observations can guarantee the validity of that law. But this view also makes it clear that probability is not simply the expression of experience. When I say: Gauss's law describes the probable distribution of the balls, I do not mean: the actual distribution of the balls comes closer and closer to this limit; what I mean is: Gauss's law is the ideal against which the degree of approximation of the actual distributions is judged. When a different distribution comes about in a particular case, we do not say: Gauss's law stands refuted; what we say is: there is an anomaly here; circumstances have combined 'by chance' to produce a different distribution.

The analogy with geometry is now perfectly clear. The propositions of geometry do not describe our actual measurements, they are the rules according to which we interpret those measurements. In the same way, probability does not state what actually happens, it states how we judge what actually happens. And just as the laying-down of the axioms of geometry is determined by the consideration that their choice lead to laws of nature of the greatest possible simplicity, so the choice of a metric in the probability calculus is guided by considerations of utility, i.e., it should be made in such a way that the smallest possible number of distributions appear as deviations from the norm. The metric can, of course, be changed – just like geometry. But this does not mean that the probability calculus is being changed. For the choice of the metric, which is arbitrary, is to be distinguished from the probability calculus itself, whose only task is to derive other probabilities from given probabilities; only the latter, and not the application of the probability calculus to fields of experience, is here being claimed to have purely logical validity.

This view points the way out of the difficulties in which the statistical theory has become enmeshed and which make it unacceptable as a foundation for the probability calculus. The problem how we reach precise limits from a finite number of experiences simply does not arise, since probability is defined at the outset by a mathematical formula; how we can erect a mathematical theory on this formula is then no more of a problem than how we can develop one out of the axioms of geometry. Thus it is really our view and not the statistical one which leads to a clear understanding of the mathematical formalism of the probability calculus.

The correctness of our view becomes perfectly clear as soon as we turn to the applications of the probability calculus in theoretical physics and, in particular, to the hypothetically assumed molecular processes. Here the relative frequencies have not themselves been objects of observation, at least not at the time the theories were developed. On the other hand, it is well known that considerations of probability are here indispensable. The attempt, still made by Boltzmann, to derive Maxwell's law of the distribution of velocities from mechanical assumptions about the elasticity of molecules has failed. The reason for this is not difficult to see. What is really needed is a new assumption which goes beyond mechanics, namely a stipulation of a metric for the probability that a molecule is in a determinate state characterized by its direction and velocity. In the calculation, an indeterminate function is first found and then chosen in such a way that the theory is successful, i.e., that it leads to the previously known laws.

Perhaps the picture we have sketched is not yet detailed enough to produce complete conviction. Let us therefore view the actual situation from another side and revert to our example of roulette. Suppose we have found a definite probability for the location of the ball: Does this mean that this event will recur with the corresponding frequency in a large number of games? It may be that the statistical frequency agrees approximately with the probability we have calculated; we then rest content and ask no further questions. But what do we do if the statistical experience yields a different value for this frequency? In the first place, we shall look and see if we did not overlook a factor on which the occurrence of the event depends; thus we shall carefully examine the surface on which the ball rolls to see if it is uniformly smooth; we shall analyse the process of pushing to find out if this is the cause of the deviation, etc.

If we have succeeded in finding such influences, we regard the deviation as 'explained'. But if we do not succeed in finding such causes, no matter how hard we try, there is nothing to be done but to explain: the deviation was pure 'chance'. The word 'chance' only indicates that we forgo an explanation. Let us suppose that our calculation is correct, that the ball comes to rest on 'red' about as often as we had expected. What is confirmed by this trial is only this: that the occurrence of the event is *independent* of other circumstances – of which we have no further knowledge; in other words, there is no further dependence except for the laws we have adduced. *To give the measure of probability is thus to stipulate when we shall speak of chance and when not.*

True, we expect such chance occurrences to be only the exception. If they happened frequently or even as a rule, this way out would not satisfy us either. We would then have the right to demand an explanation. In the search for such an explanation, we should eventually come to the verge of altering the system of natural law. But whether we would actually do this depends ultimately on whether the new laws of nature were simple. It is impossible to draw a boundary here and say: the laws of nature reach up to here; chance begins there. It must be said clearly here that all this is not a question of fact; that it is entirely arbitrary how many of the given facts we explain or how few, and what degree of complexity we still admit in our concepts. True, the degree of complexity depends on the facts. But one can only say: the more complex the laws of nature we admit, the smaller the number of such chance occurrences; and this shows anew that it is only a matter of stipulation when we speak of 'explanation' and when of 'chance'.

It is commonly thought that the application of the probability calculus rests on a special fact which has been called the 'law of equalization by chance', the 'law of great numbers', or 'irregularity'. If nature did not have this peculiarity, then – so it is thought – it would not be possible to apply the probability calculus at all. Such equalization does in fact occur in all chance phenomena. Is it not then perfectly correct to state that there is a special fact here, that the laws of nature do not by themselves adequately describe the course of events but need to be supplemented with a further item, i.e., the law of equalization by chance? But this, I believe, is a mere misunderstanding. Above all, the law of great numbers is not a law, it simply expresses that we seek no further law. We establish a law when we come upon a regularity. The fall of the die, its reflection on the surface of the table, its rotary motion, etc. are all subject to law. When we have established these laws, there is nothing left to explain. When the six numbers turn up with approximately the same frequency as we go on throwing, we have no cause to be astonished, for we have no cause to expect anything else. We should only be surprised at a deviation from the equal distribution; only this would create the need for an explanation. If it is said: But all this does not get rid of the fact of equalization; what is astonishing is that one does not get continuous exceptions to the equal distribution, then we reply: This manner of expression is misleading; it is entirely arbitrary what is to count here as 'exception', and what as 'chance' and what not. If the improbable became the rule, if our dice acquired a sudden preference for certain areas, then we would just have to alter our laws of nature. But when we would stop saying that it was a matter of chance and begin searching for an explanation – this moment cannot be fixed precisely. Even the question itself is misleading, for it makes it appear as if the laws of nature were something independent, something confronting us objectively, whose presence or absence we could only accept as a fact. Such an idea must be thoroughly expunged. The laws of nature are only a means of representing phenomena. They are, to put it precisely, the formal structure we insert between the statements which describe our actual observations. When we speak of a law of nature and when not depends exclusively on the utility which the introduction of such a formal structure affords us. True, we cannot know whether there are simple laws of nature; but this we know in advance: if there are laws of nature, then the law of great numbers must also be valid, for it is only a different expression for the fact that we have now reached the point where we stop searching for laws of nature; that we are satisfied with the level of explanation we have reached. The law of great numbers, which has not ceased to exercise the imaginations of philosophers since the days of Bernoulli, has no longer anything mysterious for us. It is the simple consequence of the existence of the laws of nature. This connection has perhaps always been felt; perhaps the sense of wonder at the existence of the laws of nature has only been expressed unclearly as wonder at the existence of the law of great numbers. But once this connection has dawned on one, the law has also ceased to be a problem for one; instead, one will see clearly that there is only a pseudo-question here. When we have reached that point, our knowledge is complete; the discovery of new facts can in no way alter the probability we had calculated.

In setting forth this view of probability, we have not really expressed any new ideas but merely formulated clearly and distinctly what has always been felt obscurely: that probability is *more* than a mere record of frequency and yet, that it has something to do with that frequency. But a clear picture of the true relation between these concepts has been a long time in the making. The statistical view has placed a one-sided stress on the element of frequency and has therefore neglected the connection between frequency and the other facts. The definition of probability as the degree of possibility, which has now been abandoned, has at least had a true feeling for the logical element in this concept, but it could not do justice to the significance of statistical experience since it took the degree of possibility to be something given a priori. Only by taking.both elements into account can we reach a satisfactory elucidation.

NOTES

[1] I do not know how far my views agree in detail with Wittgenstein's. Wittgenstein is preparing a major work which will also deal with the concept of probability.

[2] I owe the following considerations to Wittgenstein.

[3] And more specifically, with a syntax containing the possibility of unlimited precision.

[4] Cf. L. Wittgenstein, *Tractatus Logico-Philosophicus*, 5. 12ff.

[5] '$p \vee q$' means 'p or q, or both', and '$p \cdot q$' 'p and q'.

[6] J. M. Keynes, *A Treatise on Probability*, 1921.

[7] I am deferring the detailed proof of this assertion. In the meantime, cf. Th. Radaković, 'Die Axiome der Elementargeometrie und der Aussagenkalkül', *Monatshefte für Mathematik und Physik*, 1929.

[8] It is clear that the probability calculus cannot teach us anything about whether this relation will also hold in the future. We do not know this, and if we assume it, then it is a supposition founded on induction just like any other. To examine the justification for our supposition cannot be the task of the probability calculus.

CHAPTER III

THE CONCEPT OF IDENTITY*

The concept of identity gives rise to various questions.[1] Is identity a relation? Is it a relation between things? Or is it a relation between the names of things? Does the law of identity say that each thing is identical with itself? Or does it require that a name be used always in the same sense? To get clear about these questions, let us inquire into the use of language, i.e., in this case, into the use of the word 'the same'.

How do we use this word? Let us look at some examples:

(1) We say: 'The man who is now entering the room is the *same* as the one I saw previously in the street'. What is meant here by the expression 'the same'? Well, what counts here as the criterion for its really being the same man? Is it his clothing, his build, his appearance, the colour of his hair, etc.? None of this is decisive; for it is certainly conceivable that there exist two men who resemble each other 'to a hair' but are nevertheless not identical. To find the criterion we are looking for, we only need to ask ourselves how we could tell the difference in the latter case. Well, obviously by their being in different places at the same time. The criterion of identity is thus *continuous existence* in space. It is in this and only in this sense that we employ the word when we want to establish, e.g., the identity of a criminal with a certain person: we follow the criminal's track back into the past to see where it leads to. Or to take another example: when shots are fired from several rifles and we examine the projectiles that get stuck, we can ask: Which rifles do they come from? That is, is this projectile here the same as the one that was previously in that barrel?

(2) But what if I say: 'Sing the same note!'? What counts here as the criterion of identity? The comparison with memory or with a paradigm.

(3) But we could also take a note to mean a note that was sounded today in this room at such-and-such a time and lasted for fifteen seconds. We can then say: Different people heard the same note. But it would be nonsense to say: 'Sing the same note!' If we take a note to mean something that extends

* First published in German in *Erkenntnis* 6 (1936) 56–64. Reprinted in Reitzig.

through time and has a beginning and an end, then the grammar of the name of a note is again more similar to (but not the same as) the grammar of the name of a person.

(4) To create a situation similar to that of notes, we would only need to agree that 'Mr. N.' was to mean not the person himself but the appearance of the person; e.g., upon seeing a second person with a very similar appearance, we would have to say that it was also 'a Mr. N'. And in fact, children do sometimes employ words in this manner.

Let us now compare these examples. All of them contain the expression 'the one is the same as the other'. What this means depends in each case on what we regard as the criterion for it. With corporeal things it is spatial continuity: with notes it is something else, and in case (2) it is again something else than in case (3); and the word 'the same' has, accordingly, a different meaning in each of these cases.

Once we have taken notice of this fact, we soon become aware that the expression 'the same' is used with an enormous number of different meanings. Compare, e.g., these cases: 'This is the same way we took at the time', 'I take the same train into town every day', 'He has the same build as I', 'Japan has been ruled by the same dynasty for three thousand years', 'I get up at the same time every day', 'Granite has the same hardness as quartz', 'The church is built in the same style as the castle', 'Here you have made the same mistake again', 'He has the same right to be here as you', 'Both have taken the same examination'.

The meaning of 'the same' is sometimes unclear, blurred, in the sense that we do not quite know whether or not to use the word. Is a church that has been restored after being half destroyed by fire still the same church? Is a wave that runs across the beach and overturns still the same wave? Am I still the same person as when I was a boy? The only right answer to this is: Say what you will. The question 'Is this object still the same' can be understood in either of two senses: (1) in the sense of the question whether we still want to speak here of the same object, i.e., *what is to be regarded as the criterion of identity,* and (2) whether this is really the same object, i.e., *whether the criterion of identity is satisfied.* The first question is answered by an arbitrary determination on our part, the second by experience.

We are now ready to take issue with a view which is responsible for a peculiar unclarity in modern logic. This is the view that the different meanings of the word 'the same', as displayed in our examples, fail to capture what is

meant by identity in the strict sense of the word. We thus read in Carnap:
"In the ordinary use of language, as well as in its scientific use, identity is *not
taken in its strict sense.* Objects which are not identical in the strict logical
sense are treated by linguistic convention as identical".[2] It thus seems as if
besides the various meanings of the word 'the same' there was yet another
as it were purer concept of identity and that it alone was of concern to logic.
It is easy enough to understand what leads to such a view. We may have
told ourselves that two notes are never really the same note; they are at best
similar notes. And is the man who is now entering the room really the very
same man as the one I saw previously in the street? Will he not have
changed a little in the meantime? Are we not quite right in wanting to speak
only of 'generic identity' in such a case?

Such accounts mistake the essence of the matter. When it is said 'Two
notes are never the same note', this is perfectly correct. For a note is then
taken to mean something that is sounded at a certain time, and the note's
not being broken off is counted as the criterion of identity. But if we choose
a different criterion – which we are free to do – then we can, of course, say
'That is the same note', and not just in a loose but in a perfectly strict sense.
It all depends on what is to be understood by 'the same note'. We have no
right at all to distinguish between a strict logical concept of identity and a
loose one; for the concept 'the same' is strictly defined in both cases. The argument proceeds as if there was somehow a pre-eminent sense of 'the same'
in which things were really the same. But there is, of course, no question of
such a sense: identical is what we define as identical.

Now what is supposed to be the nature of this 'logical' identity? We are
told that two things are identical when they agree in all their properties, i.e.,
when they cannot be distinguished. This is Leibniz's famous 'identity of in-
discernibles', which Frege and Russell introduced into modern logic. Well,
the word 'identical' can be defined in this way; only it must be said that the
resulting concept has nothing to do with the ordinary meaning of the word
'identical'. It is not true at all that two criminals are identical when they cannot be distinguished. That is not how we use the word 'identical'.

But let us look at the matter more closely. Two men are supposed to be
identical when they cannot be distinguished; and this definition, it is intimated, holds strictly speaking only for a moment; for a moment later the
man has already changed a little. Well, let that pass; but what shall I do to
test whether a man is identical with himself at this moment? Shall I try to

distinguish him from himself? How do I go about that? The difficulty lies evidently in the word 'can be distinguished'. What does this mean? Does it mean that I have made the attempt to distinguish him from himself and that the attempt has failed? Or does it mean that I could not even make the attempt? Russell's exposition makes it appear as if we could make the attempt and as if it was only the failure of the attempt which established that the two persons were identical. To make this clear, let us take a look at the use of the word 'distinguishable'. We do not say: 'The armchair over there cannot be distinguished from itself'; we say: 'The one armchair cannot be distinguished from the other', i.e., the two armchairs are perfectly alike; they have the same colour, the same form, the same size, etc. It will be objected that this example only confirms Frege's definition; for they can then at least be distinguished by the place they occupy; and the definition says that they are identical when they agree in *all* their properties, and hence, when they cannot be distinguished at all. But this is nothing but a misunderstanding. For if there was no such thing as the fact of impenetrability in our world, if it was therefore possible to put the two armchairs in exactly the same part of space so as to make them coincide, then we could not say, as long as we held on to our language: 'They can no longer be distinguished'; we would not speak of two armchairs at all. This talk loses its sense here, whereas it made perfect sense to say of two armchairs in different parts of the room that they could not be distinguished.

Let us simplify the example. Instead of the two armchairs, let us imagine two circles of the same colour. In viewing the circular images, I shall perhaps say: 'I notice no difference; i.e., I see the same colour, form, and size', or perhaps: 'I notice a difference, e.g., in colour'. Let us now imagine that the two circles move ever closer together till they coincide in the end – does it still make sense to ask: 'Can I distinguish them or not?' Evidently not. What cannot be distinguished now is something quite different, viz. the circular physical discs. It can be said of them: 'I cannot distinguish them in this position'; but with respect to the images this talk has lost all sense. It can be said of the visual image: 'The two circles have the same size', 'they have the same colour', but not: 'they have the same position'; for it no longer makes any sense to speak in this case of two images.

It can be seen how the error in this way of thinking comes about. We imagine a gradual approximation which is to *end* in identity, and which proceeds like this: At first the two forms come closer together (the two armchairs become more and more similar); then, towards the end of this process, the last difference, the one in place, begins to disappear; and in the end the two coincide completely: they have become indistinguishable and hence one. We thus think that in the course of this process they have become less and less distinguishable, and we conceive of identity as the limit of distinguishability. But in reality it is like this: If it makes sense to ask whether the armchairs can be distinguished, then they are two armchairs; if this question makes no sense, then it is one armchair. In other words, the question whether two things are identical[3] is not the question whether they can be distinguished, but whether *it makes sense to ask* whether they can be distinguished.

The following considerations should throw light on the matter from another side. A word has meaning only in a sentence. The meaning of a word is characterized by the way in which the word is combined with others. We therefore ask: how do *we* use the word 'the same', and how does Russell use it? This brings out at once an important difference. We say for example: 'The projectile lodged here is the same as the one that was previously in that barrel' or 'Paul is the same person as the one who was previously called Saul'. We thus place the word 'the same' either between two descriptions or between a description and a proper name.[4] But we never say 'Paul and Saul are the same', except in the sense that the words 'Paul' and 'Saul' are used in the same way; and this is not a statement (not a description of a fact) but a rule which allows us to replace the one name by the other. Russell, on the other hand, places the sign of identity between two proper names: he uses it to form the sentence $a = b$. He thus seems to want to say something about things, not about the names of things, and in this way he deviates from linguistic usage. If he had said: By '$a = b$' I mean that 'a' and 'b' are signs for the same thing, this would be unobjectionable. Instead, he defines the sign of identity in such a way that the expression '$a = b$' stands for a statement, viz. the statement 'a and b agree in all their properties'. We already know what led him to this definition: it was the opinion that the identity of two things could be shown in experience through a comparison of their properties. Russell seems to have overlooked entirely that, as a consequence of his definition, the proposition 'a and b agree in all their properties' becomes a tautology, so that it does not express a fact of experience at all. For if that proposition is supposed to say that a and b are not two things but only one thing, then this surely means that 'a' and 'b' are different signs for the same object; which means in turn that they are synonymous, intersubstitutable; but if the signs 'a' and 'b' are synonymous, then the proposition 'a and b agree in all their properties' says exactly the same as the proposition 'a and a agree in all their properties'. So Russell either means by 'identity' something *entirely different* from what we mean by it; in which case he cannot use the symbol 'a = b' in the same way we do, viz. to express the fact that 'a' and 'b' mean the same. Or he uses it in our sense, in which case he must admit that the sign 'a = b' is a rule which cannot as such be confronted with experience, and his definition becomes useless. The reason for this whole confusion is that Russell expects the same symbol to carry out two different tasks, first, to reproduce an experience, and secondly, to give expression to a rule.

Someone might perhaps try to defend the definition by saying: But do we not use the locutions 'a and b are identical' and 'a and b agree in all their properties' under exactly the same conditions? Wherever we employ the former, the latter is also true (for it is a tautology); and where the former is inadmissible because we are dealing with two different things, the latter, too, becomes false. It follows that Russell's definition expresses precisely what we mean when we say that a and b are identical.

The reply is that reflection on precisely this objection makes the misunderstanding stand out even more clearly. When we are dealing with two different things, the proposition 'a and b agree in all their properties' in Russell's sense says something false; in spite of this it is, of course, an empirical proposition. But it follows from this that it cannot be the negation of the proposition 'a is identical with b'; for this is a tautology. That is: if we assume Russell's definition, the propositions 'a and b are identical' and 'a and b are different' do not stand in the relation of affirmation and negation – and this does not, evidently, correspond to the sense in which we use those words.

But what, then, is the opposite of the case that a and b agree in all their properties? Well, it is that they differ in at least one property. When experience shows that two crystals agree in all their mechanical, optical, and chemical properties, the opposite of this case is that they differ, e.g., in their optical behaviour. In both cases I speak, of course, of *two* things. But when I have only *one* crystal in front of me and refer to it at one time as 'a' and at another as 'b', then it is *logically* excluded that a and b differ, and the proposition 'a = b', interpreted in Russell's sense, is now a tautology, so it no

longer represents the opposite of the case that a and b have different properties. A definition which does not bring the concept 'identical' into opposition with the concept 'different' does not, obviously, reflect the meanings
of these words.

In conclusion, it may be well to take a look at the consequences of this
conceptual confusion; we may then see the more clearly how false interpretations and false problems cannot even arise. If we use Russell's sign of identity in the sense of our word 'identical' – and this is Russell's intention –
then we are tempted to think that a formula like '$(\exists x) .x = x$' expresses that
there are things, that 'there is something rather than nothing'. (And it is
easy to form a series of analogous formulas.) But such a formula says, in
fact, no more than can be inferred from the definitions of its signs. If it is
transformed according to Russell's definition, then it means simply that
there exists an x which has all its properties in common with itself. But this
is a tautology; it says nothing at all, and hence, it does not say that 'there are
things'.

We must therefore distinguish very sharply between the formulas in Russell's notation and the interpretation of these formulas. The formula '$(\exists x).
x = x$' can, of course, be formed, and it is not our intention to represent it as
contrary to the rules or senseless; only it does not say what it seems to be
saying – a proof of how easy it is to be misled by that notation. And the reason is precisely that it only *seems* to agree with our linguistic usage.

Let us discuss one last objection to our account. It could be said – and
this objection was raised by Carnap – that our account leaves it unclear
whether or not the propositional function '$x = y$' is to be admitted; if it is,
then one can by substitution form the proposition '$a = b$', which we have
declared to be inadmissible; if it is not, then it is unintelligible how we can
arrive at the propositional form '$a = R'b$'. We thus get either more propositions than we want to admit or not all propositions.

Our reply is this: We can form the propositional function '$x = y$', provided we are clear about what it means. If I want to represent the proposition 'Paul is the same person who was previously called Saul' by '$a = R'b$'
in the notation, then the sign '$=$' has the same grammar as the word 'the
same' – and hence *not* Russell's grammar. But if we take Russell's function
'$x = y$' and replace the variables by the values a and $R'b$, then we also seem
to get the statement $a = R'b$; but it only seems so, for what this statement
means is that a and $R'b$ agree in all their properties and not that a *is* the ob-

ject $R'b$. The meaning of the latter proposition depends on the criterion by which we test whether, e.g., Paul and Saul are really one and the same person.[5] And we have seen that, while this criterion is different in different cases, it never consists in this: that the two things agree in all their properties. It can be seen how the objection falls apart: It is not true at all that the proposition $a = R'b$ (where the sign '=' has Russell's grammar) expresses what we mean in everyday language by 'a *is* $R'b$'; the latter proposition does not result by substitution from Russell's function '$x = y$'; and this disposes of the presupposition from which the objection proceeded.

NOTES

[1] For valuable suggestions in developing the present view, the author is indebted to numerous conversations with Mr. Ludwig Wittgenstein concerning, among other things, the concept of identity.

[2] R. Carnap, *The Logical Structure of The World*, § 159 [The passage is here given in Mr. Kaal's translation.]

[3] Identical in the sense of: present only once.

[4] We here assume that 'Paul' is a proper name; whereas in fact the name 'Paul' stands for a description; but this does not matter for our purposes.

[5] Here it comes out that 'Paul' stands for a description.

MORITZ SCHLICK'S SIGNIFICANCE FOR PHILOSOPHY*

The need for a 'renewal of philosophy' is voiced with increasing force in the history of philosophy. Great thinkers have always refused to take the ideas of their predecessors for granted. Rather, they have always tried to penetrate to the ultimate foundations of knowledge. Thus Descartes, Locke and Kant each felt they were acting as turning points and initiating a new way of philosophical thinking. That feeling was right; for the philosophical mind has made through them a step forward which can never be reversed.

A survey of the last eighty years of philosophy reveals an image as confused and distraught as it never was before. Some, depressed by the collapse of the great metaphysical systems, pin their faith on a return to Kant. Others try to build up a philosophical world-picture out of the results of science. Science should not be our guide, yet others claim, the true source of philosophy is inner experience; only inner experience leads us to the depths of being. Still others maintain that the basis from which any philosophical speculation has to start is not experience but acts of consciousness directed to the understanding of meanings. And then there are yet others who assert that philosophy does not in any way proclaim objective truths but is to be considered as the mere expression of a personality, of a psychological type. It has been said, twisting round a statement by Zola: 'Philosophy is a temperament, seen through a world-picture'. And it has been thought that philosophers have actually nothing else to do than to catalogue and to 'einfühlen' what mind created in the past.

We are thus confronted with a number of opinions, each of which claims to possess the truth. All this, no doubt, is the symptom of a deep-rooted crisis. Such a picture has already been drawn so often that it is useless to dwell on it; but we are throwing some light on it here in order to emphasize better

---

* First published in Dutch in *Synthese* 1 (1936), 361–70. The present translation is based on one by Philippe van Parys and Arnold Burms. Waismann used some of this material in the opening pages of the work posthumously printed as *Principles of Linguistic Philosophy,* 1965, and the translation from German there used has also been drawn on. The German original appears to be lost.

the fact, unnoticed by most, that philosophy has now already reached a decisive turning point. Obviously, only a small circle of people qualified to judge can grasp the exact import of this turn. But it is beyond any doubt that the future lies with the new way of thinking. We are referring to the ideas originated in Frege's and Russell's works and now adopted by Wittgenstein, Carnap, the group of Polish logicians and various American trends. One of the main representatives of this style of thinking is Moritz Schlick, around whom the 'Wiener Kreis' gathered. Schlick's philosophy is the principal attempt ever made to overcome the splitting up of philosophical systems by building up a conception which does not belong to any direction but sets a norm to all directions. The following lines are devoted to this new way of philosophizing.

Attention has so far been paid almost exclusively to the answers given to philosophical questions. Discussion has up to now been concerned with these answers, their being right or wrong, their justification or refutation. The new conception already departs from the one nearly always accepted so far by completely abstracting from the answers and carefully investigating the nature of the question itself. When one penetrates behind the verbal formulation of a philosophical question and goes back to what gives rise to the question, one is faced with that kind of perplexity with which all philosophers, from Plato to Schopenhauer, were engrossed. When Augustine wonders at memory, when the validity of the mathematics becomes a puzzle for Kant, then there is a strange kind of uneasiness emanating from the questions. Everybody who has tormented his mind with a philosophical question knows by experience that he soon reaches a point where so to speak everything is foggy for the eye of his mind. The perplexity underlying these questions ultimately arises from a deep-rooted lack of clarity. What we are seeking in philosophy is to get rid of this lack of clarity. But what is this liberation supposed to consist of? Does it consist in the acquisition of knowledge? Are we not dragged again into the whirlpool of opinions? That the solution has to be found elsewhere is suggested by the following argument. When dealing with a philosophical question, one is always held captive by an analogy which equates the philosophical problem with a scientific one. Here as there, it is argued, the solution of a problem is to consist in knowledge which can be expressed and communicated in a sentence. This conception is too narrow. When e.g. the problem of flying has been discussed, the solution consisted in a technical construction, towards which knowledge is merely a guide. Who would think that the solution of the social problem consists in a theoretical insight? Let us note that people also talk about the solution of the problem of movements in modern painting. If one considers this series of examples, one discovers that solutions are of different natures: they sometimes consist in knowledge, sometimes in the production of some work, sometimes in an act that changes reality, etc.

The philosophers' great mistake was that they gave answers before examining the nature of their questions. It did not occur to them at all that the mere form of the question might already involve a fallacy. Therefore, they did not really descend to the roots from which the uneasiness arises; they contented themselves with pseudo-solutions, which may have dazzled temporarily by their fascinating brilliance, but which did not prove satisfactory in the long run.

Remarkably enough, the impulse for change came from a completely different side. In the hans of mathematics, logic had stealthily grown into an intrument which by its subtlety and expressive power went far beyond the fossilized logic of the schools. Though originally made for the purpose of analysing mathematical inferences, the new logic did not remain restricted to that field.

Frege made a remark which now sounds prophetic. This logic, he thought, might once be useful to philosophy, the task of which is 'to break the tyranny of words over mind', by discovering the fallacies which almost inevitably arise from the use of verbal language. What was surmised by Frege has been actualized by Wittgenstein and Russell. They used the resources of the new logic to elucidate the structure of our language. What emerged was that the linguistic form often conceals the logical structure of our thoughts. These investigations led to a much clearer insight into the nature of a philosophical question: such problems torment us when we do not understand the nature of our language, so that we labour under the delusion that we are investigating a substantive problem while we are just misguided by the way we express ourselves. The greatest danger is that this seduction, without our being aware of it, gets hold of us and captures us again and again in linguistic traps, under the form of countless images, metaphors and similarities.

This insight leads to a radically new solution of philosophical problems, very different from what one expected. For one had expected definite possitive or negative answers to questions such as whether the external world actually exists or whether the human will is free. But instead of giving such answers, the more penetrating analysis leads to the insight that the form of those questions rests on a misunderstanding. Analysis thus gives us a feeling of liberation by elucidating the meanings of oùr words and their combinations so thoroughly that we are freed from the need of asking any such question. Once this point is reached, one realizes with amazement that most of these questions disappear without trouble (unless they change into scientific questions) – the uneasiness has suddenly vanished. In a remark he left behind, Schlick put the key idea of his philosophy as follows:

In former times philosophy asked about the origin of Being, the existence of God, the immortality and freedom of the soul, the meaning of the world and the rules of action – we do not ask anything but: what do you mean actually? To everybody, whoever he is, whatever he is speaking about, we ask: what are the meanings of your words? Most people are taken aback by it. But it is not our fault, we ask this question in all honesty and do not try to set a trap for anybody.

Now, how does one discover 'the meanings of words'? How can we help somebody else to understand the meanings of his words? We bring him to think about the meanings of words, i.e. carefully to call up before his mind the rules according to which he uses these words. We all know such rules on the level of grammar. If it is allowed to extend this formulation to all rules, including the countless conventions of language, then we may say that the meaning pointed out is a fact in logical grammar. *We* do not construct this logical grammar. We rather ask others: How do you use this word? In this or that way? We provide him with several such rules, and in choosing one of them he fastens down a piece of logical grammar, i.e. we just draw his attention to what he is actually doing and abstain from asserting anything. There is here no room for disagreement, for disagreement can only arise when something is asserted and thus challenge by an opponent roused.

We are now beginning to see what the change consists in: it is a turn to an objective, undogmatic philosophy, which can have no opponents since it does not assert anything. Philosophizing in this spirit is just thinking with clarity and lucidity.

This method radically changes the situation of philosophy. First of all it emerges now that the so-called unsolvable problems – and these the genuinely philosophical ones – are pseudo-problems. For that they were unsolvable was our own fault, since we have built questions out of unallowed combinations of words and have thus stepped outside the 'Sinnbereich' or 'sense-realm' of language. Every question which can meaningfully be asked can in principle be answered; for giving the meaning of a question is nothing but giving the conditions under which it is to be answered by yes or no, and this is already a step towards the answer. The system of these unsolvable questions is what from of old has been called metaphysics. This metaphysics has had its day, 'not because the performance of its task is an activity which goes beyond the capacities of human thinking (as e.g. Kant thought), but because this task does not exist. Laying bare the wrong way of questioning also makes the history of metaphysical controversies intelligible.' (Schlick, 'Die Wende der Philosophie', *Erkenntnis,* 1930)[1].

The concept of logical grammar outlined here is nowadays acquiring an increasing significance. Obviously one must not think of a grammar which seems arbitrary and contingent. Looking more deeply, one discovers beneath this surface a completely different kind of rule, which confers sense on our linguistic combinations. Philosophy now has the task of laying bare this whole complicated organism of rules that is still partly hidden at present, and of making us clearly aware of them. The perspective thus broadens and leads to a general philosophical grammar. It is the true actualisation of the mathesis universalis, which was Leibniz' dream. The great currents of thought of the past flow towards it. The whole of mathematics, the rules of logical inference, the system of conventions that pervades our physics and to which Poincaré has drawn our attention: all these are merely parts of one universal grammar. Completely new problems here turn up at the horizon of philosophical thinking. Where do the rules of logical grammar come from? Is here any secret agreement between them and reality? Do they reflect the essence of reason? Or are they arbitrary, so that alternative rules could be substituted? The investigation of these questions reveals that the logical relations of our language stand out in a very specific way against a background of open possibilities. If one examines these possibilities, one is led to completely new and surprising perspectives – e.g. to conceivable experiences which cannot be actualized in our world but the description of which sheds some light on connections of our actual experience which, we think, can only be conceived in one way. Only then can we understand by what contingencies the formation of our concepts has been determined so far, and also how other experiences would steer the formation of our concepts into other channels. We can in this way conceive for instance of experiences which could lead to a completely new use of the word 'I'. Schlick used this method very fruitfully in one of his last essays, 'Ueber die Beziehungen zwischen physikalischen und psychologischen Begriffen' (in *Revue de Synthèse,* Paris, 1935)[1].

The latter insights contrast with several ideas Schlick has argued for in his still important *General Theory of Knowledge*[2]. This provides us with an example of the intellectual courage with which Schlick disengaged himself from his former opinions out of love for new insights. The first part of his book, containing some fundamental insights in the nature of knowledge based on a precise knowledge of mathematics and physics, is still unexcelled in its kind. Also the treatment of the philosophical problems of space and time which form the background of Einstein's relativity theory – and particularly his criticism of classical philosophy, in which sharp light is shed on the shortcomings of Kant's theory of knowledge – belong to Schlick's main achievements. In the field of philosophy of nature, Schlick was an undisputed authority; in explaining statements and concepts of the most recent and difficult physical theories (quantum theory), he was unequalled. His expositions of present-day controversial issues relating to the boundaries between physics and biology brought his pupils into touch with modern research. And in his seminar he spread about rich suggestions concerned with a great variety of subjects: illusion, 'Ganzheit', 'Gestalt', the concept of consciousness, that of the self, the mind-body problem, the foundations of arithmetic, the analysis of the concept of probability, of that of cause, and much more.

Schlick, however, was not only a severe and relentless thinker; the great problems of life were perhaps still closer to his heart. Already his small book *Lebensweisheit*, which he started writing while still a student, deals with ethical problems. And questions about ethics and culture occupied his mind during the last years. Asking these questions manifested another, more personal side of his being. And now something has to be said which at first seems surprising: there was something deeply childlike in his personality; a powerful and indestructible faith in the purity of youth pervaded it. Nobody else has celebrated the cheerful happiness of youth in such glowing terms; the enthusiasm which sets our hearts aglow for a cause, an action or a person and the enthusiasm of youth are the same fire. This trait is a trait of genius. Every really great mind has the spotless innocence of a child. What is light, cheerful, clear and innocent – that is what Schlick summarized in the one word 'youth'. In conformity with Nietzsche, Schlick claims that life is meaningless as long as it is completely dominated by purpose; being as such, existing as such has no value – it must also have a content and its meaning completely lies in it. Freedom from the tyranny of purpose can only be reached in those activities that we perform out of mere joy. To the extent that these activities are most condensed in youth, it may be said that youth is the meaning of life.

The notion of disinterested youth leads to the notion of goodness, which Schlick puts at the core of his ethics. A man acts rightly if he aims at the good out of inner joy and not if he abides by an externally imposed rule of duty. Schlick succeeded in justifying this idea against other views by strictly scientific means in his *Problems of Ethics*[3]. The conclusion of this work is a hymn – in which he consciously imitates Kant's hymn to duty – an apotheosis of his highest ethical concept:

Goodness, great and dear is thy name! Thou hast no harshness in thee that would exact respect without love. Thou entreatest men and they follow thee. Thou threatenest not. Thou hast no need to lay down laws. Nay, of thyself thou findest way into men's hearts and they revere thee willingly. Thou smilest and all thy sister desires lay down their arms. So glorious art thou that we need not ask whence thou comest. Be thy descent what it may, thou hast ennobled it.

This intellectual attitude springs from Schlick's personality, which also included a poetic touch. If one wants to understand Schlick, one should not overlook this feature. He did not belong to those enlightened characters who rejoice at the removal of the pressure exerted by great metaphysical questions. Schlick once drew a distinction between genuine and journalistic metaphysics. He had a deep respect for the spirit which emanates from the metaphysical systems of the past, as he had for anything great. Yet, he considered it a mistake when the world feeling which expresses itself in those systems wraps itself in the robe of pseudo-scientific knowledge. When asked about Schlick's attitude towards metaphysics, one can answer that it completely depends on which meaning is attached to this word. If one defines it as a system of propositions which seemingly present a kind of physics, and hence something like a world description which does not exactly know what it describes, then it should be said that Schlick has rejected it, precisely because he demands rigour and clarity, from knowledge. If, on the other hand, one defines metaphysics as a fundamental feeling of life– and that is no doubt the much more essential meaning of the word–as an openness to the seriousness and the meaning of life, then it is really the keynote of Schlick's whole philosophy.

Schlick's main characteristic is his love of truth. He could not bear mere brilliance. In spite of his tolerant way of thinking, Schlick was on the whole critical and easily saw the weakness of an argument. 'Rechthaberei' and dogmatism were foreign to him. He had an undestructible optimism, due to his conviction that goodness and justice would ultimately triumph. This was also experienced by his numerous pupils, attracted by the clarity of his ideas, charmed by the harmony and moved by the goodness of his personality.

NOTES

[1] [It is planned to have two volumes of the Vienna Circle Collection devoted to Schlick's papers. Those here referred to will be included, in English translation.]

[2] [New York, 1974, a translation of *Allgemeine Erkenntnislehre*[2] Berlin, 1925.]

[3] [New York, 1939, an authorized translation, by Mr. D. Rynin, of *Fragen der Ethik,* Vienna 1930. The passage quoted, however, has been newly translated for the present volume.]

CHAPTER V

HYPOTHESES*

1. HYPOTHESES IN SCIENCE

What is an hypothesis? Let us start our investigation in science, where the nature of an hypothesis is to be met with in its purest form. Imagine a physicist who observes that the electrical conductivity of a metal varies with its temperature. To follow up this dependence, he systematically alters the conditions of the phenomenon in an experiment. He first subjects the metal to different temperatures, observes the corresponding conductivity by the deflection of a needle, notes the numerical values, and enters them as points in a system of co-ordinates. He then looks for a law which reproduces all the observed values – or to put it in concrete terms, a curve which joins the individual points. Once he has succeeded in setting up such a law – *how* he finds it, whether by rational deliberation or by intuition or in some other way, will not now be discussed – he can use it to predict the behaviour of the metal under different conditions that have not yet been investigated. The law thus goes beyond the observations: it claims to hold also for those cases that have not yet been observed, and it calls on us to test this claim. This function of a law of nature is so conspicuous that we could even say: it is a recipe for making predictions.

But what does such a prediction state, and what does such a test really establish? It is always the occurrence of a certain fact of experience, like a pointer indicating a certain point on a scale as in our example. When the experiment furnishes the experiences we had been led to expect, we say that the hypothesis has been *confirmed* (or *corroborated*); otherwise we say that it has *run foul* of the facts. In more precise terms this would mean: the hypothesis is the starting-point for deductions. Its application consists in the derivation of its consequences and their comparison with the findings of observation.

Needless to say, a proposition which describes an observation ('the poin-

---

* Composed in German, probably before 1936, as a chapter for *Logik, Sprache, Philosophie* and now printed (Stuttgart 1976) as an appendix to that work.

ter indicates point 5') is not in turn a recipe for making predictions. If we try
to put this difference in words, we might say: we have here two contrasting
kinds of statement. One kind describes what may be called non-recurring
experiences. The other kind undertakes to fit these experiences into a lawlike context, to fill the gaps, as it were, left by particular observations.
Propositions of the latter kind are hypotheses.

Physics constructs a system of hypotheses in the form of a system of
equations. A physical hypothesis has sometimes been conceived as if it were
only a summary of prior observations, something like a table in a compend-
ious and manageable form. But this conception misses precisely what is
most characteristic of physics, i.e., the reference to the future. Physics is not
history; it makes prophecies. A physical law refers to the future ad infini-
tum. Unlike a statement which describes a non-recurring experience, it
never counts as certain: we always reserve the right to drop it or to alter it as
soon as new experiences occur that are not in accord with it.

Although this account is perfectly correct, it contains the hidden danger
of a misunderstanding, to which we now want to call attention. It might be
thought that an hypothesis is a proposition just like any other, except for
one difference: it has not yet been tested in all cases, so that we are less certain of its truth. But if the mark of a complete check is regarded as the distinguishing characteristic, then the difference is expressed in a misleading
way: It sounds as if the concept of checking (of running through all individual cases) had a sense in the case of an hypothesis, and as if this check were
impossible only for technical reasons (which presupposes that the attempt
is conceivable). Let us imagine that I want to find out whether a road is
lined with nothing but acacias. After inspecting only some of the trees, I
say: 'So far it's true; I don't know whether it will continue to hold true, but I
suppose it will'. Here the word 'suppose' has a good sense because it is used
in opposition to 'know'. But what if the road is infinite? What would it
mean to make a supposition about *all* trees? What am I now distinguishing
a supposition from? When we notice this difference, we are inclined to say:
An hypothesis is an *eternal* supposition. But here we are, so to speak, placing the emphasis again in the wrong place. The point is not that there is always some uncertainty attaching to an hypothesis, but how an hypothesis is
used. And there cannot be the least doubt about its use: an hypothesis
serves to predict future experiences – and again only *concrete* experiences.
It is the connecting link we insert between the actual experiences of the past

and the actual experiences of the future. Indeed, to an astronomer it never really matters whether the law of gravity 'is valid for all eternity' – this would be a question which we could never come any closer to answering, no matter how many observations we made – but whether it *stands the test*, whether he *succeeds* with this assumption. It could be said: *The sense of an hypothesis is the work it does.* An hypothesis behaves therefore very differently from a proposition which has only been checked incompletely. The words 'complete' and 'incomplete check' are extremely misleading. An hypothesis is, in truth, a different grammatical formation.

But would it not then be better to distinguish an hypothesis from propositions proper and to say that an hypothesis was only a recipe for forming propositions? If we take a proposition to mean something which is used according to the rules of logic, i.e., which can be affirmed and denied, from which we can draw inferences, etc., then an hypothesis is a proposition. It is only a different kind of proposition from observation sentences, and it is the desire to stress this difference which moves us to attempt such a formulation. If we want to characterize the place of an hypothesis, we would do best to describe its grammar. We would then have to say:

An hypothesis does not follow from a small finite set of particular propositions (i.e., propositions which describe non-recurring experiences). It is – in this sense – never verified. On the other hand, an infinite number of particular propositions can be gathered from it. A more accurate description of the relationship will have to be postponed till we have extended our field of vision somewhat.

II. HYPOTHESES IN DAILY LIFE

The totality of what we see and touch does not yet form a lawlike self-contained world. We supplement what we perceive by fitting in the unperceived, thus filling in the big gaps between appearances. We act like an archaeologist who reconstructs an ancient temple from a few remaining ruins, by piecing out the fragments, retracing the lines, and putting in the missing ties in thought. These supplements to our sense perceptions will be called hypotheses in an extended sense of the word. We make use of such hypotheses at every step. When we hear a rumbling noise and say that a vehicle is passing in the street, when we see an arm waving from a window and instinctively attach a man to it in thought, we fit the perceived phenomena into a thought context. It is characteristic of an hypothesis that future experiences can tell for or against it, that it carries with it an element of uncertainty.

Hypotheses make their appearance in the mere description of the physical world. But extreme care is required if we are not to lapse into certain closely connected errors. Thus it would be incorrect to say that the proposition 'Here is a die' is an hypothesis in the sense we sketched above,[1] but it has at any rate some *similarity* to the hypothesis used in the description of an aspect. An hypothesis here is what connects the individual perceptions; it is as it were the curve we draw through individual points. And this holds for all descriptions of physical objects. It might now be objected: what if I am deceived by an hallucination or by the fog or the diffraction of the light? In this case the construction would only become more complicated; but even then it would ultimately be a matter of fitting the propositions corresponding to the aspects of the die into a certain system – only into a more' complicated one.

To designate such a fitting, language makes use of the expression 'it seems'. 'I saw a lamp on the table', as it must be understood in our ordinary language, says more than the description of the visual image. 'It seems that there is a lamp on the table', while expressing something phenomenal (an appearance), indicates at the same time some uncertainty. It sounds as if it did not describe anything real, but something whose nature was still unclear, but in fact 'it seems' only says that something is being described as a special case of a general rule, and what is uncertain is only whether further experiences can be described as special cases of the same rule. It is characteristic of the word 'it seems' that we cannot say 'it seems to seem'.

'It seems' here characterizes what is not hypothetical: the non-recurring experience; on the other hand we say 'A stick in water seems to be bent' and express by this that it is not really bent. And this ambiguous use of the word 'it seems', the voicing-along-with-it of the thought 'but it only seems so', indicates perhaps a certain tendency of our language to neglect the experience of the moment in favour of the permanent reality of the physical world. (And the reason for this lies, of course, in the different importance of the two descriptions.)

The grammatical relationship between the words 'to be' and 'to seem' is, however, more complicated than might be supposed after our first schematic account. Thus it must be said that the question 'Does there seem to be

a man there?' comes to no more than the question 'Is there a man there?', for
it does not contain a description of the visual image. And if I ask someone
'Is there a man there?' and he replies 'Yes, there seems to be a man there' –
can I reproach him for not having answered my question? Hardly. On the
other hand, his answer might also have been 'There is no man there',
though not 'There seems to be no man there', for I would then be justified in
saying: 'Don't you know it for sure? Take another look'. The words 'to be'
and 'to seem' are thus often used as opposites, but sometimes also in a parallel way.

III. VERIFICATION IN DAILY LIFE.

A proposition like 'Julius Caesar crossed the Alps' is also an hypothesis.
Asked how it is verified, we could point to different things: ancient texts, in-
scriptions on monuments, etc. But what do future experiences have to do
with the proposition? It is directly connected with the sense of the proposition about Caesar that we may conceivably find Caesar's corpse, but also
that we may conceivably find a document from which it emerges that no
such man ever existed and that his existence has been made up for certain
purposes.

We cannot usually tell by looking at a proposition of everyday language
how it is to be verified. And this holds not only for propositions like the one
about Caesar just mentioned. Almost every statement of daily life provides
an example of it. Suppose I say 'There is a sofa there'. What is the verification of this? Is it that whoever enters the room receives a visual impression
of a sofa? But then we are not told where he is supposed to stand (whether
by the door or in the corner), what the image is supposed to look like, etc. In
short, the way it is to be verified is not yet contained in the wording of the
proposition. But then it is not yet determined by the proposition, but must
be laid down expressly. That is, we must make the determination that we
will regard the proposition as verified under such-and-such (to be described
in detail) circumstances.

But the important point is that we need not make such an unequivocal
determination, nor do we make one in many cases. We do not know exactly
what we still admit as a verification, and even if we are offered different
possibilities, we are unable to make a final choice between them. We are un-
decided, as in the example of 'Moses'.[2] When I say 'I put the book on the

shelf today' – can I furnish on demand an exact list of all the sources of error by which I may have been deceived? Hardly. But this means that we do not know in advance when we would take back the proposition; we have made no determination.

There is, incidentally, something peculiar about taking back a proposition. Suppose I am asked 'Is there a man out there?' and I look out and say: 'Yes, there is a man'. Under what circumstances would I actually take back such a proposition? That would be extraordinarily difficult. If I am told by way of reproach 'But there was no man, nobody saw him' – would I say 'Oh, I see, so I was mistaken'? Nothing of the sort. I would say 'What! But I saw him as large as life. You can't talk me out of that. The others must have been deceived'. I would stand by my assertion with the utmost tenacity and would not know at all under what circumstances I could be argued out of it.

Our discussion touches here on the problem of belief. Since Hume, many logicians have been concerned with induction and with the question whether it can be logically justified: whether we just believe or whether we have a *reason* for believing. But the question 'What is belief?' has usually been set aside in these investigations.

To find out what belief is, let us imagine a man who is being dragged forcibly into a fire. He will resist with all his might and lash out with his arms and legs like a madman. Is this reasoning? Is the man looking for reasons? Does he recount the number of times he has been burnt already and infer from this that it will probably turn out badly for him? No. He lashes out like a madman – well, that *is* belief[3]. If men do thousands of things every hour and, as the saying goes, build upon induction when they believe that fire burns or water quenches thirst, then belief is almost never a conscious process but consists in what men do.

Let us now return to the verification of an hypothesis. The proposition 'There is an armchair there' is an hypothesis in the sense that future experiences can tell for or against it. It is conceivable that in walking towards the armchair I run into a glass-wall and then notice that it was a mirror image I had seen; I would in this case take back the hypothesis on the basis of what is called experience, but *this experience would not in turn be expressed by an hypothesis.* The proposition 'What is confirmed by experience can be refuted by experience' holds for *hypotheses* only if it is understood as follows: What is confirmed by experience can be refuted by *another* experience.

We now come to another important point in our considerations. An hypothesis, as we use it in our ordinary language, brings together very different kinds of experience. For example, I hear someone playing the piano next door and say: 'My friend is in'. If I am now asked how I know that, I could reply: 'He told me that he would be next door at this time' or 'I hear someone playing the piano and recognize his style of playing' or 'I heard a step a while ago, which was just like his', etc. It now seems as if I had verified the same proposition each time in a different way. But this is actually not quite right. What I have verified are different facts which count as symptoms of something else. The playing of the piano, the step, etc. are symptoms of the presence of a friend.

This example is enough to show how the propositions of our language bring together very different kinds of experience. To get a clearer picture of their structure, let us imagine an hypothesis as a massive three-dimensional body. The phenomena we observe are then, as it were, the individual cross-sections we make in different places through the body. In other words, the individual experiences are inserted in the hypothesis like cross-sections in a three-dimensional body. *Strictly speaking*, what we can verify is always only *one* such cross-section. In cases where it looks as if we had verified the same proposition in different ways, we have really verified different cross-sections of the same hypothesis.

We can now extend our talk about verification to the hypothesis itself and say that it is *verified* or *falsified* through the cross-sections. We must only be clear that we are then speaking of verification in a different sense of the word.

So an hypothesis brings experiences into a lawlike context, and this is true not only of experiences at different times and in different places, but of experiences of different kinds – of radically different facts. Now it is important that not all of the experiences of seeing, hearing, etc. brought under an hypothesis need actually occur. Imagine a creature with a compound eye, with which it can see things two-dimensionally and estimate angles the way we do with our eye, and in addition with two feelers, with which it can touch objects. Now suppose this creature has certain sense experiences which it expresses in the proposition: A sphere is moving towards me. This will be an hypothesis in the sense explained: it puts together different kinds of experience. Let us now imagine the creature without the experiences of its feelers; it could still describe the whole of its experience in two-dimensional terms: as a circle in its visual field which is getting bigger. (By choosing this example we want to avoid certain complications which our mode of perception would bring along with it.) But where the experiences of its feelers are missing, the creature could also prefer to represent its experiences by means of the hypothesis of a sphere. The hypothesis would then carry with it *more* than was required for the task of describing the immediate experience. An hypothesis contains as it were an idle wheel: as long as there occur no further experiences, the wheel remains unused; it enters into action only when further experiences are brought in. An hypothesis is thus designed for *more* than the reproduction of one kind of experience (in our example: the measurement of angles without the experience of the feelers).

If we now imagine one kind of experience to be missing entirely from our world – e.g., the experience of seeing – one way in which physical-object statements can be verified will have been removed; and if we also imagine the experience of hearing to be missing, a second possible check will have ceased to apply. In each case, we are indeed confronted with 'the same' propositions – but the work they do is different in each case. The general principle involved is this: What is verified in different ways does more work than what is verified in one way. (That is, if we say that we have verified 'the same' in different ways, then 'the same' means more than what is only verified in one way.) This throws, incidentally, some light on the popular belief that the laws of physics are independent of sense perceptions. Thus it has been argued that men would have arrived at the same optics even if they had no eyes, just as they were able to investigate the physics of ultraviolet light, even though they lacked a specific organ for those rays. It may well be true that men would have arrived at the same laws of nature, i.e., at the same mathematical equations; but there is more to be said. An equation acquires a physical sense only when we are given a method for co-ordinating the numerical values of the equation with the propositions of our language, so that we can make the transition from a mathematical equation to observation sentences. Now if these propositions have a different character (as in the world of the blind), then the sense of the physical equation has changed accordingly. We can now do fewer things with them, e.g., we can no longer use them to make predictions in the way we do, and this means that the work they do is different.

IV. CAN THE PROPOSITIONS OF DAILY LIFE BE CONCLUSIVELY VERIFIED?[3]

[Not reproduced in this volume. See footnote]

V. IS AN HYPOTHESIS MERELY PROBABLE?

Since in a large class of cases an hypothesis cannot be conclusively verified, the question arises whether we can still speak of it as either true or false. If I can never know whether an hypothesis is verified, can I still call it true? Let us consider what is meant by this question. If we speak of truth and falsity in the case of an hypothesis, we must at least, on pain of giving rise to a mis-understanding, hold on to the rule

$$p \text{ is true} = p, \ p \text{ is false} = \ \sim p.$$

The question whether we can call an hypothesis true would thus amount to the question whether we can assert an hypothesis as such, and hence, whether we can say for example: 'There is a ball here?' Of course we can, and this is really all that needs to be said.

But an hypothesis is surely verified in a very different sense from a singular proposition – should we not therefore say that it is also true in a different sense? Have the primitive concepts of logic not perhaps changed their sense here? To say that the words 'true' and 'false' had a different meaning would be to say that they were used differently, but is this the case? According to our linguistic usage, 'It is false that there is a ball here' says exactly the same as 'There is no ball here'. The question whether the word 'false' does not perhaps have a different meaning would thus amount to the question whether negation means something different in this case. This is an indication that the question is not yet quite clear. For what is supposed to be the criterion for saying that the word 'not' does not have the same sense as in the other case? If the criterion says that $\sim\sim p=p$, that $\sim\sim\sim p= \sim p$, etc., then the sense is the same. But if it says that what follows the 'not' is used according to the same rules, then there is a difference, i.e., precisely, the difference between an hypothesis and a singular proposition. The expression 'description of the verification' may well have a different meaning in the two cases, but not the words 'true' and 'false', which are only component parts of the T-F notation, just like the words 'not' and 'or' by which they can be replaced.

But would it not be better to call an hypothesis neither true nor false but only probable? Let us consider what the consequences of this terminology would be. According to it, we could not say: 'There is a ball here', but only: 'It is probable that there is a ball here'. But if that was all we could say, then it would again make no sense to speak of probability. Our insistence on calling an hypothesis probable would only be justified if the concepts 'true' and 'probable' were used side by side (in the same system), and hence, if the proposition 'It is probable that there is a ball here' was *distinguished* from the proposition 'It is true that there is a ball here'. But the latter proposition was supposed to be excluded by that proposal. What, then, is the concept of probability supposed to be distinguished from?

The proposition 'We cannot speak of truth, only of probability' could also be taken to mean that we ought to say 'probable' instead of 'true', to distinguish the case of an hypothesis from others. This word would therefore characterize a proposition as an hypothesis. But this is not how the word 'probable' is used. It is used in *opposition* to 'true', not as an analogue to the word 'true' in a different region of grammar. But if we wanted to give the word 'probable' a new meaning, so that it served to make an hypothesis more conspicuous, this would already be impracticable for the reason that the phrase 'it is probable that...' would have to be prefixed to *every* hypothesis and hence to the overwhelming majority of propositions in our language, in which case this prefix could just as well be omitted. All that is right about this proposal is that an hypothesis must be distinguished from a proposition in the narrower sense.

The same applies to the remark 'We can never know whether this is a ball'. If what is meant here is a logical impossibility (in contrast to the case where we can try to find out whether this is a ball and our attempt has failed), then this proposition can only be intended to distinguish the grammar of an hypothesis from the grammar of another proposition. The remark 'We can never know' would then be a grammatical remark.

VI. CAN AN HYPOTHESIS ONLY BE DECIDED IN ONE WAY?

We now come to a question, the answer to which will at last give us a complete insight into the structure of an hypothesis: What is the relation of an hypothesis to experience? Or more precisely: What is the grammatical rela-

tion of an hypothesis to the propositions in which a physicist records his experimental findings? According to the received opinion, the observation
sentences *follow* from the hypothesis. The business of verifying consists in
drawing the consequences of a law of nature and confronting them with
experience. Since a law of nature has infinitely many consequences, the
ideal of complete verification is unattainable, whereas on the other side a
single negative instance seems to be sufficient to cause the downfall of the
law. A fact of great significance would follow from this: while a law of nature could never be strictly confirmed, it could be strictly refuted; or, as we
might also say: a law of nature could only decided in one way. Does this
view accord with scientific practice?

There are, no doubt, cases where everyone would agree that a theory was
refuted by *one* observation. If a law of nature was put forward tentatively,
if it was given little chance even at the outset because of other considerations, and if it then failed the test the first time it was confronted with experience, it would be dropped without hesitation. It is quite different if what is
in question is a well-confirmed and generally accepted law, which may even
form the corner-stone of an entire theory. What astronomer would give up
Kepler's laws on the basis of *one* observation? If some deviation were actually observed, we would first resort to various other possible explanations
(observational error, deviation of the planet due to unknown heavy masses,
other kinds of disturbances, e.g., due to friction with thin gases, etc. etc.),
and only if the resulting structure of hypotheses found too little support in
experience, if it now became too complicated and arbitrary, if it no longer
satisfied our craving for simplicity and clarity, or if we were offered a better
theory, would we decide to drop those laws. And even then, the refutation
would not be definitive, valid for all times: it could always turn out that
some circumstance had eluded us which would make the whole appear in a
different light when it was taken into account. The history of science records a number of cases where the apparent defeat of a theory was turned
into a complete victory (Olaf Römer, Leverrier). Since we cannot look into
the future, we have no guarantee that such a situation will not recur.

Anyone who looks at the facts without preconceptions will, I believe,
agree with the conclusion we are now drawing: that a theory is never,
strictly speaking, refuted by one observation. The true relationship is much
more complicated, and it is only adumbrated in these words. The example
just mentioned, that of a generally accepted theory coming into conflict

with a single observation, makes us realize the length and complexity of the chain of ideas leading from an observation (like the coincidence of a point of light with a cross-wire) to a statement of the theory. In case of conflict between theory and observation, we must first test the strength of the whole chain link by link before we can accept the observation as trustworthy. But this test is only possible if the observation is repeatable, and this shows already that a single observation proves next to nothing. There will always be links in the chain which, when one of them breaks, will render all observation useless.

This is the general view of things. In a particular case there may be reasons for a different decision, but in principle the situation remains always the same: A single observation is incapable of refuting an hypothesis.

We had therefore better formulate the relationship more carefully, by saying that certain observations (or more precisely: observation sentences) tell *for* or *against* an hypothesis, which does not mean that they confirm or refute it. Instead of this we could also say that an observation fits well or badly into a general law. How much value we attach to a contrary observation, and when we regard it as a 'refutation' of the theory (in the practical sense), depends on the whole scientific situation of the case, and it appears to be a hopeless task to set up precise rules here.

The decisive point is this: A single observation can never exclude an hypothesis in the sense in which the negation of p excludes the proposition p. There always remains the possibility of making the observation accord with the hypothesis by introducing further assumptions. In other words: If hypothesis H leads us to expect observation p, but what we actually observe is $\sim p$, then $H . \sim p$ *never represents a contradiction.* But if we were dealing, as is commonly thought, with a strict logical inference, then the denial of the consequent would entail the denial of the antecedent. But there is never any question of this; on the contrary, it always makes sense to say: The hypothesis holds, even though this particular observation tells against it.

It will now be said: This is because we have oversimplified the connection. The observation p does not really follow from H alone, but from H and a series of further presuppositions which are often not fully articulated. If the expected observation p does not occur, this just means that one of the other assumptions is false.

Our reply is this: This view would be correct if the system of presuppositions could be stated precisely and exhaustively in each particular case. But is this the case? Can we ever be sure that we know all real conditions on which the outcome of even the simplest experiment depends? Evidently not; what is formulated is only *part* of the conditions – the ones that are known to us, on which our will has some influence, the ones we can survey – the others merge into an undifferentiated mass, into the vague presupposition that there exists 'a normal situation', that there are no 'disturbing circumstances', or whatever other turn of phrase we may choose to indicate the possibility that unforeseen circumstances may come into play. The relation of a hypothesis to particular propositions is therefore strictly speaking as follows: If such-and-such laws of nature hold and such-and-such limiting conditions obtain and there are otherwise *no disturbing influences*, then so-and-so will come to pass. And here it must be pointed out emphatically that the italicized words contain a hidden presupposition which cannot be analysed into clear particular propositions. This 'presupposition' is never used in the actual derivation of an observation sentence, it never enters as a premise into a chain of inference – but in that case it should not be called a presupposition at all; for a premise from which no inference is drawn is not a premise. In fact, those words only say that in case of conflict between theory and observation we will *look* for disturbing circumstances, but that we reserve the right to hold on to the theory.

We can, of course, imagine the 'ideal case' where we can survey all conditions in the world completely. We must only realize that we are then constructing a grammatical model which does not correspond to reality; the concrete scientific situation is always different. The schema does not thereby lose its value. All it really does is emphasize the *aspect* under which we would like to look at the situation. And the task of the philosopher is only to point out *that* it is an aspect.

It is therefore meaningless to say: If all conditions in the world were known to us, all particular propositions would follow in all strictness. Since we have no prospect of gaining a complete knowledge of all conditions, the fact remains that a particular proposition is not a compelling consequence of general laws. It seems to us that it amounts to the same thing whether we say that the system of conditions is incomplete and that the inference is therefore lacking in stringency, or whether we say that there is from the start no strict logical inferential relation between the *known* conditions and a particular proposition.

From all this it appears that the connection between hypothesis and ob-

servation is looser than has been imagined up to now. We could, incidentally, arrange the propositions of our language in certain strata, by admitting into the same stratum all those propositions between which there exist
precisely formulable logical relations. Thus the propositions of mechanics
or thermodynamics can be arranged in a system whose elements stand in
precisely statable relations, and in which it can always be decided strictly,
for any two propositions, whether one is a consequence of the other,
whether they contradict each other, etc. The statements made by an experimental physicist in describing certain observed data, like the position of a
pointer in his apparatus, are also related to one another in precisely formulable ways. (If a pointer indicates point 3 on a scale, it cannot possibly indicate point 5; there exists here a strict relation of exclusion.)

On the other hand, a proposition of theoretical physics can never come
into strict logical conflict with an observation sentence, and this
means that there are no precisely formulable relations between the two
kinds of propositions. All the relations and connections which the logical
calculus places at our disposal hold only as long as we move within a given
stratum of propositions. But the real problem begins where two propositional strata as it were border on one another, and it is the problems of this
plane of cleavage which today merit the attention of logicians.

It is not our purpose to go more fully into these things. But even from
these very few remarks it should be clear that a proposition of theoretical
physics cannot in any way be conceived as a summary of observation sentences, as a general statement. For this would imply the possibility of *strict*
verification or *strict* falsification. And the same situation seems to recur
when we follow up the connection between statements about bodies and
propositions about perceptions. The logical form of a statement about
bodies comes out in its use, and its use is not such as to allow us to replace
the proposition 'Here is a table' by a statement dealing only with certain
ideas of sight or feel or touch. It may well be true that the former proposition is somehow 'founded' on the latter: that in justifying the proposition
'There is a table' we must point in the end to certain perceptions; but in formulating this foundation we must proceed again with extreme care. Let us
remember how difficult and complicated it may be under certain circumstances to distinguish between reality and illusion, if we really try to formulate the relationship in exact terms. None of the attempts that has been
made so far to give an exact logical analysis of the concept of a body has

reached its goal. The schema of truth-functions, even if we add the formation of general and existential statements to it, is not wide enough to encompass the actual wealth of propositional forms. Instead, we feel impelled
towards the more liberal view that there are *different types* of proposition
in our language, which cannot be produced out of *one* kind of proposition –
e.g., observation sentences – by means of certain logical building materials.

VII. HYPOTHESIS AND CONVENTION

If an hypothesis can be neither strictly confirmed nor strictly refuted, if the
scale of the balance on which experiences are weighed comes down only
gradually on the one side or the other, then we enjoy a certain measure of
freedom in determining the moment when we will give up an hypothesis or
when we will still hold on to it. I have used the phrase 'a certain measure' deliberately, to emphasize that our freedom of decision lies only within certain bounds. For in many cases we would regard an hypothesis as confirmed beyond doubt, and in certain other cases we would reject it just as
firmly.

This element of freedom which resides in every hypothesis has led the ad-
vocates of conventionalism to the view that laws of nature are nothing but
arbitrary conventions. The law of inertia and the law of the conservation of
energy in particular have been adduced as evidences for this view. We can
now understand how such an opinion could arise. Since a certain measure
of arbitrariness inheres in every law of nature, laws as such have been declared to be arbitrary. But to put it in this form is to exaggerate and distort
the idea – an indication of how easily the urge to simplify and schematize
our grammar can lead to the most momentous errors.

Conventionalism is a partially justified reaction against an older somewhat naive view, according to which the laws of nature were supposed to
reproduce the facts as faithfully as a photograph. It used to be thought that
the truth or falsity of an hypothesis could be decided unequivocally
through experience. The knowing mind stood to reality in a purely passive,
receptive relation, like a polished mirror, which reproduces things in a pure
and undistorted way.

As against this, the true state of affairs seems to be this: An hypothesis is
certainly not a spitting image or slavish imitation of experience. In it resides
something of the bold inventiveness of mathematics, that 'free creation of

our mind'. On the other hand, it is, of course, connected with the facts of observation, though this connection is represented in an unfortunate way when it is said that an hypothesis is confirmed or refuted by experience. It would be better to say that it is a frame into which observations fit well or badly, and that in judging this fit we always have a certain latitude. There is in fact something of the arbitrariness of a convention about an hypothesis, as well as something of the restrictiveness of a genuinely empirical proposition.

If the view sketched here is correct, then the opposition between 'empirical' and 'a priori' loses some of its significance. This opposition does, of course, exist, and we are not here trying to deny its existence; it only seems to us that it has been exaggerated in the past. A law of nature is not as empirical as has been imagined, nor is a convention as *a priori*. Our acceptance of a law of nature is strongly influenced by considerations of simplicity, utility, and the aesthetic form of the whole system; on the other hand, we are not free in the choice of a convention but guided by experience. It now appears that the concepts 'a priori' and 'a posteriori' stem from an earlier phase of philosophy and are no longer very suitable for giving an accurate picture of logical relations.

A very simple example should make it clear that there resides a certain element of freedom within an hypothesis. Suppose we are looking at a curve which is supposed to connect a number of points – have we already determined in advance when this representation will still satisfy us and when we will give it up? (We are disregarding the case where *all* points lie 'exactly' on the curve; instead we allow for minor deviations.) We could have anticipated all possibilities by asking ourselves: 'What would we do if the point came to lie there? And what if it lay there?' So we can either make a prior decision, so that we are, so to speak, ready for anything. Or we can say: 'I don't know what I will do in that case'. We then, as it were, take things as they come and make a decision when the need arises.

We could, of course, imagine the ideal case where we state what exactly we will regard as a confirmation and what as a refutation, i.e., where we say in advance: If so-and-so happens, I will give up the hypothesis. For under certain circumstances I would indeed give it up; and if I will give it up, I can determine the exact conditions in advance. But in reality, people are at a loss when asked to do this. They just will not let themselves be guided by strict rules. And this is what we may call the element of freedom. In other cases, we can perhaps determine when exactly an hypothesis will no longer satisfy us, but this only shows that this is a somewhat different kind of hypothesis from the other kind.

The conclusion that it is not always determined when an hypothesis will be given up places an hypothesis in a kind of twilight. There is much that is blurred in physics as well as in geometry. A convention about representation sounds often like an hypothesis, and conversely.

Let me try to explain this by first taking a fictitious example. Suppose we observe that a celestial body describes a circular orbit. If we later notice that it deviates slightly from the circular orbit, we are inclined to represent its orbit as a circular orbit with deviations (and such a representation may be useful for certain purposes). In any case, the proposition 'The body describes a circular orbit with deviations' is supposed to record a certain experience. On the other hand, it is clear that any orbit can be conceived as a circular orbit with deviations, provided we interpret the words 'with deviations' with sufficient latitude. We may therefore, if we like, hold on to this statement *for ever*; and if we do, the proposition which was originally supposed to describe an experience has now become a *form of representation:* it sketches out the schema, so to speak, to which every description of the orbit must conform. If we now forget that we are looking at a form of representation, we will still regard the proposition as if it described an experience; it thus appears to us in a peculiar twilight: on the one hand, it looks as if it was supported by certain experiences, and on the other hand, it confronts us with the appearance of irrevocability.

The example chosen was a fictitious one. If we turn to actual science, we often find the same indeterminacy, the same hovering between experiential propositions and arbitrary stipulations as shown in our example – only in a more disguised form. Let us take Hertz's mechanics as an example. This system of ideas is governed by a single fundamental law from which the whole content of mechanics can be derived. This fundamental law, a kind of generalization of the usual law of inertia, says: The natural movements of a free material system are such that the system follows a straight course with a constant velocity. Expressed in the usual terminology of mechanics, and after an explanation of that terminology, this law represents a clear and simple statement. The difficulties begin only when we follow up the application of this law to real processes. For so-called free material systems it represents an experiential proposition, or more precisely an hypothesis, to which the actually observed movements of such a system conform. The law is then really only a transformation of Newton's principles, and the same experiences which tell for Newton's mechanics also tell for Hertz's. But physics also recognizes 'unfree systems' (e.g., those in which temperature appears as the cause of motion). These systems do not at first conform to the fundamental principle; to make them subject to it, Hertz resorts to the artifice of conceiving them as parts of free systems: to tangible bodies with their visible movements he adds 'hidden' movements of other invisible masses, in such a way that the systems which their supplements conform again to the fundamental law. But here the sense of the fundamental law begins to change without its being noticed. For if we are *always* free to add such supplements, the fundamental law becomes a mere form of representation. It would then be best expressed as a prescription saying: 'Supplement the system with hidden masses and their hidden movements in such a way that the law of motion for the whole assumes such-and-such a form', which is like saying in the case of a geometrical problem: 'Construct a circle corresponding to this ellipse which...'

But the fundamental law can also be interpreted in another way. We only need to determine when the introduction of hidden masses is permissible and when not, by appealing to certain experiences; the fundamental law, combined with the assumption of hidden masses, then acquires a clear physical sense. This is at least the sense in which Hertz himself wanted it to be understood.

There are accordingly two possibilities: If we are *always* free to introduce hidden masses into our considerations, then this assumption is an idle wheel, which turns without advancing our understanding of the facts. But if we make the introduction of hidden masses dependent on empirical conditions, then the fundamental law acquires a well-defined physical content. It all depends on how that assumption is interpreted, i.e., which claims the hidden masses must satisfy, besides the one demand the fundamental law imposes on them.

Let us return to our example of circular motion. I can say 'The body describes a circular motion'; I can also say 'The body describes a circular motion with such-and-such deviations'; but it is meaningless to say 'The body describes a circular motion with deviations' if this is supposed to mean: with *any* deviations. But the grammar of the system also contains the proposition: 'Put all motions in the form of a circular motion with deviations'.

The expression 'circular motion with deviations' may occur in the grammar but not in the description.

An hypothesis resembles a curve we draw through a number of points. Asked 'What is a curve? It is the prediction that *all* points will lie on it? Or is it a determination of the form of representation?', we can only reply that it may be both. If we use it in such a way that we say 'If I find a point lying outside it, I will give up the curve', then it is a *prediction*. I have then given a criterion which would contradict it. But if I give no such criterion, if I say 'Whatever happens, I always represent the points by this curve', then I have given a *rule of representation*; I have decided to let the curve represent every point. In many cases it is *unclear* how we use a curve. We then play an 'unstructured game': we do not state when we will give up the curve, but we also do not say that we will hold on to it under all circumstances. (Think of Gauss's error curve: What is experience here, and what is stipulation?) A rule of grammar does not thereby become a statement of fact, nor does the one shade into the other; all that is required here is a stipulation to what extent the curve is to agree with reality, and whether it is to agree with it at all.

The same thing must be said of the network set up by mechanics to reproduce the facts. The net alone does not yet say anything about the facts, but only the net together with the rules for its use. We can then distinguish three cases:

(1) I say: Reality comes within such-and-such a degree of proximity to my net.

(2) Whatever reality looks like, I will always describe it with a net of such-and-such a form.

(3) I leave it open: I do not say whether I will give up the net, or I say that I will perhaps give it up, but without naming the condition under which I would do so.

Let us pursue this indeterminacy, this wavering between empirical proposition and rule of representation, by taking a further example. What about the proposition 'Space is three-dimensional'? As it stands, it seems to express a fact. We should like to say in support of this proposition: Well, whatever body I take, it has length, breadth, and height. Or we point out that at any point we can draw three straight lines perpendicular to one another. But at the same time we feel that the proposition has an altogether different character from the proposition 'There are only three precious metals'; and we will then perhaps want to see in it the expression of necessity. But what kind of proposition is it that temperature is one-dimensional? Does it express an experience? Could it be refuted? What the proposition means is that *one* numerical value will do in this case. And in the same way, the three-dimensionality of space says that the position of a point in space can be described by three numerical values. What it expresses is therefore only a rule about the form of representation. Incidentally, this triad of co-ordinates appears already in ordinary language in the equivalent use of the expressions 'length', 'breadth', and 'height', or 'above – below', 'left – right', 'forward – backward', and in the absence of a fourth analogous pair of concepts.

That we are really dealing only with a rule here can also be shown in the following way. Suppose someone tells us: 'A fly is crawling on the edge perpendicular to these two edges' (while pointing with his hand at two edges), then this statement will be true or false. But if he says instead 'A fly is crawling on the edge perpendicular to these three edges', then this is senseless (and not false). With this form of words he has stepped outside the form of representation and violated a *rule*.

But the proposition 'Space is three-dimensional' could also express an experience. Imagine the following case: The bodies in our environment disappear, and some time later, bodies with very similar properties emerge at different points in space. Someone now puts forward the hypothesis that a body has moved out of our space, described a trajectory through four-dimensional space, and then returned to ours. If he succeeded in correctly predicting the re-emergence of an object on the basis of its assumed trajectory, we should have to admit that the hypothesis had been confirmed. But it cannot be emphasized enough that we are not compelled to make such an assumption; it is merely advantageous to use such a representation. We can, if we like, stick to our three-dimensional representation by saying: The bodies have simply disappeared at one point in our space and re-emerged at another.

The proposition about the three-dimensionality of space would express an experience if we had stipulated in advance that we would take it back under such-and-such circumstances, e.g., if cases of disappearing and re-emerging objects were actually to occur. The proposition, so interpreted, would say that such experiences do not in fact occur. On the other hand, in so far as it speaks of *bodies*, it presupposes three-dimensionality in the ordinary grammatical sense. It can now be seen how easy it is for unclarities to arise: The same form of words belongs at the same time to two different systems, say geometry and physics, and if we are not clear about this, we get the confusing impression that we are stipulating that space is three-dimensional, while at the same time investigating whether it is really three-dimensional.

VIII. REMARKS ON VERIFIABILITY

Whenever we are confronted with a proposition which is in principle unverifiable, accurate reflection would show us that we had attached no meaning to one term or another. It is our own fault if the proposition has become senseless: we have combined familiar words in such a way that they have lost their familiar sense in the process.

We have said: We understand the sense of a proposition only if we know under what circumstances the proposition is true or false. We now see that we must express ourselves more precisely: The circumstances mentioned in our criterion must always be capable of being *observed* and shown. If someone wanted to say: 'The proposition "space has expanded twofold" is true if space has expanded twofold – those are the circumstances that verify the proposition', we would not accept this – and not on the basis of some philosophical theory which will only admit empirical circumstances, but because we have seen that this is a necessary condition for signs as such to acquire meaning. The proposition which describes the verification of another proposition must contain words whose use can be explained by showing or contextually; otherwise we do not understand what it means.

Thus when we say that the sense of a proposition is clear to us only when we can give the circumstances under which it turns out to be true, the word 'circumstances' can only mean circumstances that can be shown, observed, experienced.

Let us consider the following proposition: 'If Napoleon had dictated peace-terms to the Czar of Russia in 1812, the whole history of Europe would have taken such-and-such a turn'. Here we might say: Such a proposition cannot be verified; we cannot turn back the wheel of history to find out what would have happened in that case. Yet we understand quite clearly what this proposition says, and we might even venture to take sides for or against it. Therefore, an unverifiable proposition has a sense.

To find out what such a proposition says, let us attend to what would incline us to accept it or reject it. We soon notice that we rely here on certain general experiences which we have acquired in studying history. Someone who thinks that Napoleon's victory would have led to the unification of Europe under French rule could marshal all sorts of reasons: the political situation at the time, the psychological make-up of leading personalities, the effect which the defeat of Russia and the existence of a formidable army must have had on France and the rest of Europe, etc. And someone who disputed this view would call our attention to further circumstances whose effect we had not included in our calculations. In short, when we weigh the reasons pro and con in such a case and finally decide for one view, we apply the whole of our knowledge of the historical situation and of general psychological laws to the special case before us. And the truth or falsity of such an assertion depends on whether we have drawn an accurate picture of the forces operative at the time.

Does the proposition therefore say what would show that it was justified? In that case, we would have to let the justification teach us the sense; whereas we understand the sense before we know the justification. No; our language hints at a connection between two events without defining the connection more precisely (this is why the justification remains open) and states at the same time that these events have not occurred.

NOTES

[1] [*The Principles of Linguistic Philosophy,* p. 73, see p. 121 of *Logik, Sprache, Philosophie,* (the German original).]

[2] [Cf. *op. cit.,* Ch. IV, 1, pp. 69 ff.]

[3] [The example (which also occurs in Ch. XIII below) is drawn from a dictation of Wittgenstein's of *ca.* 1931: '*Diktat für Schlick*'.]

[4] [For the text of this section, see *op. cit.,* Ch. IV, 2, pp. 74 ff.]

CHAPTER VI

IS LOGIC A DEDUCTIVE THEORY?*

The concept of a proposition can be defined with the help of the logical calculus by saying: A proposition is a linguistic expression which is used according to the rules of the logical calculus. The logical calculus, however, gives rise to a number of questions, one of which will be discussed here in some detail. The logical calculus has been presented outwardly in different ways. In what follows, we choose Russell's and Whitehead's presentation, so as to have a definite example before our eyes. (But it should be said at once that the following discussion is independent of the choice of this particular presentation.)

The structure of the calculus is this: Certain formulas, the axioms, constitute the point of departure of the system: they are placed at its head without proof; and other propositions are derived from the basic propositions according to definite instructions. The logical calculus is thus presented to us in the guise of a deductive theory, whose propositions are linked by proofs, and it reminds us in this respect of other deductive systems, like mechanics or geometry. But this comparison is misleading at a decisive point, and it has in fact misled some. For this way of presenting it makes it look as if the logical calculus were a theory which progressed from certain true propositions to other true propositions, and hence, as if it were a structure of *propositions* – but how, then, can it be used to define the concept of a proposition? This is our question, and the answer to it will place the calculus at last in a clearer light.

Let us stay for a moment with the usual view. In creating their systems, Frege and Russell had in mind the following idea: The basic propositions of logic are truths which force themselves upon the intellect with absolute necessity; as for other propositions, their truth is recognized as soon as they are reduced to those initial propositions by a chain of proofs. These authors were, then, of the opinion that in seeking out the consequences they were penetrating deeper and deeper into a realm of eternal truths. That this view

---

* First published in German in *Erkenntnis* 7 (1938), 274–281 and 375. Reprinted in Reitzig.

misses the heart of the matter is shown by the following simple consideration: Logic serves to bring the rules of inference into systematic form. These
rules justify inferences. That is, I can say 'this inference is valid because I
have proceeded according to such-and-such a rule', but not 'this inference is
valid because such-and-such a proposition is true, i.e., because things are so-
and-so'. Let us put it this way: If logic were a deductive system like mechanics, then we would not really be interested in the axioms at all, but simply
and solely in the fact that the rest followed from the axioms. But in that
case, Russell's theory would not be a logic at all, but only an *example* of a
logic. For in this system we are only told that so-and-so follows from these
axioms. But surely Russell wanted to set up rules which would justify inferences – and not just inferences from his basic propositions, but inferences
as such. Once we can see more clearly, it must appear that logic is not an ed-
ifice of truths at all but only the expression of a *rule of inference,* and that
the whole structure of the system of formulas is subordinate to this purpose.

We begin the clarification by pointing out that in Russell's logic two
things are to be distinguished sharply: first, the formulas of the calculus
(the axioms and theorems) expressed in the signs of the notation, and secondly, the instructions which tell us what we are to do with the formulas;
these instructions are expressed in ordinary language. Above all, the transition from one formula to another must be reduced to precise rules. Now
the most important rule in Russell's system says; we may make the transition from a formula p to another formula q when $p \supset q$ is a valid formula
of the calculus. This rule will be referred to hereafter as instruction *I*. It represents *the* transition in the calculus which is the schematic representation
of our inferences. Thus if I infer proposition q from proposition p, my inference can only proceed according to the fixed schema

$$p$$
$$p \supset q$$
$$q.$$

And here we come to speak of the source of the unclarity. For if we look at
this schema, we get the impression as if we had inferred the conclusion from
two premises, namely the original proposition p *and* the further premise $p
\supset q$; and since the latter is supposed to be a valid formula of the calculus,
and hence an axiom or the consequence of an axiom, the upshot seems to

be that in inferring q from p we had not only presupposed p but in addition the axioms of logic. It might be thought: q follows from p only *if* the axioms of logic are true.

Thus what is misleading about Russell's logic is that it makes it appear as if our inferences required certain propositions, namely the axioms, in addition to the rule of inference *I*. But the axioms do not really figure as premises at all, but as parts of a *rule*. This becomes clear as soon as we consider how the axioms are *used* in the calculus. If we want to infer proposition q from proposition p, we must complete the schema

$$p$$
$$\ldots\ldots$$
$$q$$

by inserting a structure of the form $p \supset q$ so as to get the original inference schema. Now the task of the axioms is to supply the missing links – the pieces, as it were, that are to be fitted into a fixed frame – and thus to give the inference the normal form outlined by rule *I*. It can be seen what part the axioms play: they are not premises *from* which we infer, but means, devices, which only serve to give our inferences a certain form. The truth or falsity of the axioms does not matter in the least; what is essential is our *determination* concerning the axioms: that an inference is only valid if it results, according to instruction *I*, from p and the apparent premise $p \supset q$.

It follows that the insertion of the axioms and their consequences into the general schema of rule *I* results in different specific rules of inference. Thus if I want to examine whether the inference from p to $p \vee q$ is justified, I proceed by means of a rule which allows me to add the apparent premise $p . \supset .p \vee q$. The axioms thus serve to complete the one rule of inference, to fill out the empty schema of rule *I*, and only together with this schema do they yield definite instructions. As we progress within the calculus, we gain more and more new instructions, all of them built according to a single plan.

We can, incidentally, tell from Russell's logic by itself that the axioms are not assertions and that their only task is to give a definite form to our inferences, and we could tell this even if we did not yet know that the axioms were tautologies. The purpose of the axioms and their consequences is to justify inferences. To justify, e.g., the inference from p to $p \vee q$, Russell adds to the proposition p a complement, which makes the inference conform to the schema of his single rule of inference *I*. But if what is to be justified is the inference from *p* to *p*   *q* and not the inference from *p* and the *axioms* to $p \lor q$, then the addition of the complement $p \, . \supset . \, p \lor q$ can make no difference to the premise *p*. Russell would therefore have to admit that there exists the rule

$$p \, . \, (p \, . \supset . \, p \lor q) = p.$$

If we now ask in which way this complement (the insertion) is used, the answer is: never as an independent proposition, but always only as part of a logical product. The insertion cannot therefore be said to be true; we can only say that, added to the premise, it yields the premise. If we regard it as essential to a statement that it can stand on its own, then the insertion is not a statement.

Let us sum up: What we cannot tell by looking at the logical calculus is that its application is totally different from that of all similar calculi. The standard application of a calculus is to say: this is true, what follows from it is therefore also true. What is peculiar about the logical calculus is that we do not apply it at all in this way. Its application consists in this: we declare that the addition of a formula *f* of our calculus to the premise of an inference yields the premise, i.e., that there exists the rule

$$p \, . \, f = p.$$

But by this rule we declare *f* to be a tautology. In other words, the application consists in our saying: *this is a tautology*, and not in our first saying that this is a tautology and in *then applying it*. In precisely this determination lies the decisive point: *it* is what makes a system of deductive formulas into *logic*; and yet we cannot tell from looking at the usual presentation of the calculus that it has this application.

This result can also be expressed in a slightly different way. If we know the meaning of the signs '$\sim$', '$.$', '$\lor$', '$\supset$', e.g., from their representation by means of truth-tables, then we can show by means of this representation that the addition of an axiom makes no difference to a proposition, i.e., that the axiom is a tautology. But if we now assume that we do not yet know the meaning of the signs '$\sim$', '$.$', '$\lor$', '$\supset$', and that we stipulate that the addition of an axiom to a proposition shall make no difference to it, then we declare thereby that the use of the signs '$\sim$', '$\lor$', etc. shall agree with the use of these signs in the T-F calculus. The stipulation *first constitutes the meaning of these signs,* and constitutes it in such a way that it agrees with the grammar of the words 'not', 'or', etc., or alternatively, with the meaning of the primitive signs 'true' and 'false'. It follows that the axioms cannot be propositions of our language; instead, they first constitute what we call the concept of a proposition.

The extremely misleading character of Russell's logic comes out in this way: no one would believe that a change in the premises would destroy the essence of a proposition. At first sight it looks as if logic consisted of propositions which asserted something and which could therefore also be denied. But what would be the result if an axiom were replaced by its contrary? Would we arrive in this way at a non-Aristotelian logic, related to our Aristotelian logic in something like the way in which non-Euclidean geometry was related to Euclidean geometry? The answer to this question turns out to be somewhat surprising; for it appears that a change in Russell's axioms does not result in a new logic, but either in *no* logic or in our old logic.

Suppose we deny one of Russell's axioms – how does this destroy the essence of a proposition? Let us look at an example: Suppose we replace Russell's axiom

$$p . \quad p \vee q$$

by its negation

$$\sim(p . \quad . p \vee q),$$

while leaving the rest of the axioms and the rule of inference *I* unchanged. If I now want to infer $p \vee q$ from p, I can no longer insert $p . \supset .p \vee q$ into the inference schema; I must use $\sim(p . \supset . p \vee q)$. My rule of inference therefore reads

$$p . \sim(p . \supset . p \vee q) = p.$$

But in this way I have set up a *new* rule for the signs '.', '$\sim$', '$\supset$', '$\vee$', which is incompatible with the use of these signs as synonyms for the words 'and', 'not', 'if', 'or'. It is as if I wanted to introduce the rule

$$\sim\sim p = \sim p.$$

I would indeed be giving a rule for the sign '$\sim$', but a different one than for the word 'not', and so I could no longer think of '$\sim$' as the sign of negation. The situation is similar in the case of the above rule. We can indeed set it up, but in so doing we declare that the signs '$\sim$', '$\vee$', '$\supset$' can no longer be interpreted as logical particles, or the symbols p, q, r as propositions.

But this is not the only possibility. For if we are already familiar with the grammar of the signs '$\sim$', '$\vee$', etc., we can safely replace an axiom by its contrary, provided we also make the corresponding change in the grammatical rule for the use of the axiom, and we can then again draw valid inferences. Thus if we choose the formula $\sim(p \, . \, \supset \, . \, p \vee q)$ as an axiom, I only need to change the rule of inference in such a way that only the negation of that formula can be inserted in the schema; there will then exist the rule

$$p \, . \sim\sim(p \, . \, \supset \, . \, p \vee q) = p,$$

and I can work with this rule exactly as before: logic has remained the same. This shows very clearly that the axioms do not express at all what is essential to logic, unlike for example the axioms of mechanics. If the axioms were propositions, they could not first constitute the essence of a proposition. And yet one cannot help seeing them as propositions. This was the paradox which, we hope, has now been cleared up.

Russell's calculus gives us the picture of *one* logic among other possible 'logics'. This picture is misleading, because it obscures the very point at which logic differs from other deductive systems (mechanics, geometry). Only if we realize that the calculus supplies component parts of rules of inference does it become clear that a change in the calculus changes the rules of inference and with them the grammatical rules which govern what we call a 'proposition'. What we determine concerning the axioms first constitutes the meaning of the signs '$\sim$', '$\vee$', etc., and either assimilates it to the truth-functional calculus of our language or does not do so.

All this is not to say that a non-Aristotelian logic is impossible. It is only to say that we must not succumb to false ideas about the significance of a change in logic: we must not imagine that the propositions of our language would remain propositions as before, and that they would only be used according to different logical rules. (This would be just as false as the belief that our numbers would still remain the same if the laws of arithmetic were changed.) If our propositions are used according to different logical rules, i.e., in another calculus, then they are turned thereby into different formations.

In conclusion, I should like to make two further remarks:

First, the word 'logic' can, of course, be used in a more general sense, viz.

to designate all systems isomorphous with the one presented by Russell and Whitehead. The aim of our discussion can be said – by way of summary – to have been precisely this: to set out clearly what distinguishes logic in the narrower sense – the theory of inference – from other systems with the same formal structure; and hence, to set out clearly what is, so to speak, specifically logical about logic. This element resides in the peculiar kind of *application* of the logical calculus.

Secondly, I believe this insight throws light on the question whether we can also imagine a non-Aristotelian logic. In *one* sense the answer is simple: it is not too difficult to construct a system which exhibits in its formal features a certain similarity to the logic of *Principia Mathematica*. But the deeper problem, hidden behind the construction of such a formal system, is whether such a system can assume the *function* of our logic – or a similar one. We can only answer this question if we first give an exact account of the kind of way in which we apply *our* logic and then try to find a similar application for a system with a deviant structure. The axioms of such a system would again have to be construed as component parts of rules of inference – but of rules which circumscribed a different system of transitions from the one circumscribed by logic up to now. Now it does seem to be perfectly possible to formalize our everyday language and the language of science in such a way that we arrive at a somewhat different system of rules of inference – at a different logic. In this sense, our discussion could perhaps help to remove an obstacle on the way to the construction of such a logic.

The character of the laws of logic remained obscure for a long time. In their pioneering works, Frege and Russell still regarded the logical axioms as substantive premises from which the whole of logic and mathematics was supposed to follow, even though these authors employ some turns of phrase which point already in a different direction. E.L. Post[1] and L. Wittgenstein[2] then showed that the axioms of the propositional calculus were truth-functions which were true for any truth-value distributions of their arguments. For some time, people therefore spoke of the 'tautological character' of logic; as if it were this circumstance which was characteristic of logic. That view leads, however, to the following difficulty: If logic consists of propositions – how, then, can if fix the concept of a proposition? For the basic idea developed in this short paper – that the axioms of logic are *parts* of rules of inference – the author is indebted to conversations with Mr. Ludwig Wittgenstein.

NOTES

[1] 'Introduction to a General Theory of Elementary Propositions', *American Journal of Mathematics* **43**, 1921.
[2] *Tractatus Logico-Philosophicus*, London 1922.

THE RELEVANCE OF PSYCHOLOGY
TO LOGIC*

Of the various aspects of the question included in the title I will try to throw some light upon three topics[1]:

(1) The question as to whether there is any indubitable knowledge of immediate experience;

(2) The way in which mental acts like believing or doubting are relevant to logic;

(3) The misunderstandings which I think are involved in any causal theory of meaning.

I

Since the time of Descartes philosophers have always been deeply interested in the question whether any of the knowledge we possess is absolutely indubitable and forms the basis of the rest of our knowledge. It is well known that philosophers, under sceptical criticism, have retreated step by step to the position that it is only momentary experience which is beyond all doubt. When I am looking at a rose, and utter the words 'This is red,' there seems to be no possibility of questioning its truth. Even if I am dreaming, it is true that I am having an experience of redness. Many philosophers have distinguished between propositions such as this, which they take to be indubitable, and other empirical propositions, for example, about physical objects, which can be refuted by experience in the future and which can therefore be said to be of the nature of hypotheses. Is this view tenable?

The answer to this question depends partly upon what we understand by a proposition. Many philosophers hold that what a proposition is cannot be defined, because the characteristic feature in it is a specific state of mind which we have when we make a judgment. In order to clarify this question,

* Reprinted by kind permission of the Editor, from the *Aristotelian Society, Supplementary Volume* 17 (1938), 56–68.   1938 The Aristotelian Society. (Previously reprinted in H. Feigl and W. Sellars, *Readings in Philosophical Analysis,* 1949, pp. 211–221. German translation in Reitzig).

let us use a method which I think is helpful philosophically, namely, to invent games with words which throw light upon our actual language. Each of these games can be described exactly and completely, without reference to the complication of mental processes which every sentence of our language involves.

We will describe three games and draw some conclusions from their comparison.

(1) A person *A* who cannot use any language has been trained to point to a red or to a green piece of material whenever he hears the words 'red' or 'green' respectively. He learns the game by someone showing him what he is to do until he imitates it. We take his imitating the game to be the criterion of his understanding it.

If I say 'red' and he points to the green material, we say that this is wrong. But in what sense is it wrong? Is it a mistake or a slip? There can be no doubt as to the answer: it is a slip comparable to the case where someone makes a slip of the tongue, makes a slip in calculation, etc.: it is a wrong move in the game, but it is entirely different from a mistake in the sense of a mistaken belief.

(2) Now we will modify the game.

There is a lamp in the room which shows red and green at irregular times. *A* is to watch the lamp and to say which colour he sees. It is assumed that he knows nothing about the use of the words 'red' and 'green' in our ordinary language, except in this game. He learns the game in a manner similar to that in which he learned the first one.

Supposing that the lamp shows red and *A*, looking at it, says 'green' – in what sense can we speak of an error? Let us remember first that *A* has learned the use of the words 'red' and 'green' only in this game; he does not know the meaning of the words 'true' and 'false', or of such phrases as 'it is correct', 'it is incorrect'. Therefore we, who are watching him, can only ask whether *A* has used the words correctly, that is, in a way which is in accordance with the rules of the game. We cannot ask whether he expresses with the words 'red' and 'green' what is true or what is false. Analogously to the first game this, too, is a case of a slip, the only difference being that he says the wrong word, instead of pointing to the wrong colour. In this game there is no occasion for mistake in the sense of mistaken belief. *A* can use the words according to the rules; if he does not do so, he does not play the game. But within the game there is no possibility of speaking of 'true' or 'false'.

(3) Now we will describe a game in which the opposition true-false does apply. Let us suppose that *A* is to guess which colour the light will be. When the lamp lights up he can say: 'What I've guessed is correct' or 'it is incorrect'; 'it is true' or 'it is false'. And this can be said by those who play the game, not only by those who watch it.

The question before us now is this: Which are the features that account for the fact that we can speak of an error in the sense of a false opinion with regard to the third game, but not with regard to the first two games? The answer to this question will throw light upon the nature of judgment.

One might say that in the second game the words 'red' and 'green' do not express a statement. They merely are names for the colours of the light. It is only in the third game that *A* wants to describe a fact with these words, and that is why his utterance can be true or false. This answer, although it points to a real distinction, is not satisfactory, for it is precisely the nature of the difference between naming and describing which we want to get clear about.

There are at first sight two answers:

(a) The difference is psychological. When the speaker, by uttering the words 'red' or 'green', expresses an opinion or expectation, a peculiar mental process occurs. This differs from what is felt in naming something. There occurs a specific act of believing or thinking or supposing, and that is what makes the utterance a statement.

Whether or not there is such a peculiar mental act of believing we will not discuss now. But even if there is, reference to that state of mind will be of no help. Although we have not investigated the processes occurring in *A*'s mind when he is playing the third game, still we say that he makes judgments. We do not wish to maintain that there is not an experience peculiar to judgments, but we draw attention to the fact that this experience is not the criterion for our deciding whether a given case is a judgment or not; therefore, the reference to such an experience would be superfluous.

(b) In the third game, in which *A* had to guess, the words 'red' and 'green' are compared with reality, they are in accordance or in discordance with reality, and that is why they can be said to be true or false. A closer examination of this answer will show that it also is not satisfactory. In the second game as well as in the third one, the word 'red' was explained by demonstrative definition, and this explanation constitutes, as is generally said, a connection between the word 'red' and the perception of the red light. In applying the demonstrative definition to the game (2), *A* looks at the lamp and gives a name to the colour of the light – is it not correct to describe this as a comparison? We could even imagine the game (2) modified in such a way that *A* has a sheet of paper on which there are green and red patches of colour next to the written words 'green' and 'red'. Whenever the coloured light appears, he compares the light with the colours on his paper till he finds the right one, then passes across to the word written next to it, and utters the word. In the game (3) where he has to guess, he utters the word before the coloured light appears and the word is subsequently compared with reality; in the game (2) the light appears first and the word is uttered afterwards. This is the only difference I can see. The words 'red' and 'green' have as much, or, if one prefers, as little connection with reality in the second game as in the third.

The real reason for the difference must be found in the different rules for the games. And here we can notice the following difference: In the second game it is not permissible to give the wrong name; in other words, if *A*, looking at the red light says 'green', this is contrary to the convention; he violates the rules and thus ceases to play the game. In the third game, on the other hand, he remains within the rules of the game even when he guesses wrongly. There are the two possibilities of his guessing the colour and of his not guessing it, and these possibilities are distinguished within the game as 'true' and 'false'. This is what constitutes the difference, and not any process of believing or judging which might accompany uttering the words.

We do not wish to maintain that the uttering of the word is true or false *because* it is a statement, but rather that it is a statement because it *can* be true or false, that is, because both possibilities occur within the game.

We are led by considerations such as these to the formulation of what is essential to a proposition: a proposition is what can be true or false; that is, a proposition is something which obeys certain rules of a calculus of truth-values. Thus a proposition is always to be understood as embodied in such a calculus. There are many different calculuses of truth-values; hence the concept of proposition is limited in different ways according to the calculus to which it is referred. So to have an exact concept of proposition, it is necessary to specify the calculus which defines it. Ordinary language is, of course, so blurred that we cannot derive from it a clear-cut notion of proposition.

To take an analogous case: if we are asked to define a real number, we would describe the calculus of real numbers. In the same way we would define a proposition by describing a particular propositional calculus.

Let us return to the original question as to whether we possess any indubitable knowledge. It is obvious that if I say 'There is a chair in the next room,' I am asserting a proposition which may be true or false. Even if it is false, my asserting it is not excluded by the rules of my language. But now let us take the example of my looking at a coloured patch in my visual field and saying 'This is yellow.' Suppose that what I see is in fact blue. In what sense is my utterance false? We must not forget that the colour-words 'yellow', 'blue', etc. are explained by demonstrative definitions and that these definitions form part of the grammar of the colour-words. Therefore, if I say, while referring to a blue patch in my visual field, 'This is yellow', I am sinning against the rules of grammar. In this case, following the rules for the use of colour-words which can be codified in a list, I am compelled to choose the word 'blue' and I have not freedom to choose any other colour-word. This case is therefore entirely different from that of game (3), in which the player has a free choice to choose as he wishes. In game (3) there are two choices, each of which would be in accordance with the rules of the game; in our case the choice has been predetermined by the rules of grammar previously fixed. My utterance, 'This is yellow', is a falsely formed proposition, but not a false proposition.

What is the case in this example holds, I think, of every description of immediate experience. If I describe a pain, a sound, or any other experience, the only question that can be raised is whether or not I use the words correctly in accordance with the rules of language; but not whether my utterance is true or false. What, then, is the nature of the utterance? It is not a proposition, if we understand by a proposition something which can be true or false. We can say, if we like, that it is a border-line case of proposition in a rather similar way to that in which a tautology is a degenerate case. But it is perhaps clearer to say that it is not a proposition at all. It corresponds to what Schlick has called a 'Konstatierung'[2], a term which he used to describe the only synthetic propositions which cannot be doubted. We are now in a position to separate what is true from what is false in Schlick's view: I cannot doubt a 'Konstatierung', not because I am so sure of it that I cannot mistrust it, but because to doubt it does not make sense. In speaking of doubting, we must be very careful to distinguish the kind of doubt referring to the truth of the proposition from the kind of doubt referring to the correct use of words. In the case which we are considering only 'grammatical doubt' is possible.

If this analysis is correct, it shows that Mr. Russell is mistaken in thinking that "in a critical scrutiny of what passes for knowledge, the ultimate point is one where doubt is psychologically impossible." At the same time, it confirms and explains Mr. Braithwaite's view that what is in question is "something whose correction is logically impossible"[3].

Such an account as I have given removes, I think, a good deal of the mysterious air which many philosophers have attributed to descriptions of immediate experience. My account shows that there are no such things as empirical judgments which are indubitably true. Every empirical proposition can be doubted, and what cannot be doubted is not an empirical proposition.

II

In this section I want to say a little more about the problems involved in the attitude of doubt, chiefly in order to see more clearly if it is true that the study of particular mental phenomena is important for the logician. Consider a case in which a child is instructed to look at a lamp which shows successively three different colours and then to call out the names of the three different colours. What he has learnt is only the use of the colour-words, but not the use of propositions. Is the child able to doubt? The next question is: What do we mean by the word 'doubt' in such a case? Perhaps that the child hesitates, stammers or that the words are accompanied by a feeling of uncertainty? Suppose the child hesitates in uttering the word 'red'. Shall we say that the child doubts that he has seen a red colour? Or that he is doubtful as to whether he is using the right word? But the question leads us in a wrong direction. The child uttered the word 'red' and hesitated – that's all that happened. To ask whether the child's doubt refers to the fact or to the use of the word would only be correct if he were able to ask himself questions like these: 'Was the light really red?' 'Is this really the right word?' that is to say, if he had learnt to think in propositions. In what does the possibility of doubting consist? It consists in the fact that a more complicated game can be played by adults than by this child.

It is obvious that a child who has only learnt to call out the names of things, but who is not familiar with the use of propositions and the difference of true and false, is not in a position to doubt. In this sense we may say that the possibility of doubting is bound up with the language, and that this possibility appears as soon as the learning of language has reached a certain stage of development.

On the other hand, there are cases in which we do speak of an animal's doubting. A horse wades through a river, testing the depth at each step. If someone is inclined to call this behaviour 'doubt', we will not object; we would only point out that what is here meant by 'doubt' is just *defined* by this behaviour.

In the first case doubting is not a state of mind which stands as it were *behind* the words. And this brings me to an important point. The relation between the doubt as to whether or not the light was red and the proposition which expresses this doubt is not of the same nature as the relation between a toothache and the proposition which states that I have toothache. A toothache may exist without being expressed in language; but a doubt cannot exist without such an expression. For doubting expression in some form of symbolism is essential.

It is true a person can be in a mood of uncertainty; if we wanted to describe such a state of mind with the help of the verb 'to doubt', we should have to use it in an intransitive way. This use of the verb 'to doubt' is similar to the intransitive use of the verbs 'to be afraid' and 'to yearn'; for there are certainly cases where we speak of an experience of yearning without reference to an object, and then we mean by it a certain objectless feeling: '*Ich sehne mich und weiss nicht recht nach was*'. In the same way we may speak of an intransitive fearing, if we want to describe certain bodily sensations of anxiety, e.g., a feeling of constriction in the throat. But such experiences do not constitute what in ordinary life we mean by 'fearing', 'longing', 'doubting'. In the ordinary use of these verbs it is essential that they should refer to objects, that is to say, that it does not make sense to say, 'I am doubtful, but I don't know of what'.

Brentano said that it is this reference to objects which is characteristic of mental states like thinking, doubting, wishing, fearing, hoping, etc. In opposition to him, I hold that the reference of a thought, a doubt, etc., to its object is determined by language. A doubt cannot exist without a language in which it is expressed; nevertheless, to doubt something is not the same as to utter the words, 'I doubt whether...' The doubt is more than the words in which it is expressed. Let us, therefore, introduce a distinction which may be shown by the contrast of the phrases

'expression of the doubt' 'description of the doubt'

and analogously

'expression of the wish' 'description of the wish'
'expression of the fear' 'description of the fear'
etc.

Expression of the doubt is the proposition expressing the doubt; description of the doubt is the proposition describing what occurs in the mind of a person when he is doubting.

The situation then is this: 'To be doubtful of...' has not the same meaning as the phrase 'to express a doubt'. The relation between the two facts is that the expression of a doubt is part of that doubt, so that the doubt could not exist without its expression. The words expressing the doubt may be accompanied by a peculiar state of mind, say, a feeling of uncertainty, or they may be part of a characteristic form of behaviour. The description of the doubt is, therefore, composed of the expression of the doubt together with the description of certain other occurrences.

But this is by no means essential. A person may express a doubt without feeling a specific experience or without showing a characteristic form of behaviour. In such a case the words do not allude to a hidden mental state in the mind of the doubter, indeed, their utterance may be the only process which takes place.

In the case where there is a certain feeling of uncertainty behind the words, it can be said that the intensity of doubting is revealed by the modulation and timbre of the voice. That is to say, we can use the modulation of the voice to indicate the intensity of the doubt. In many cases the hesitating sound would represent the doubtful state of mind. On the other hand, this cannot be the criterion for doubt; since it is perfectly possible for a person to speak in a hesitating voice and to behave as if he were in doubt without in fact being in such a state.

In practical life we make use of various symptoms in order to diagnose whether a person merely pretends to doubt or is really doubting. But none of these symptoms can be said to be the defining criterion for doubt.

Mr. Russell in the paper of this symposium is very near to the truth when he explains that "my belief that today is Tuesday or Wednesday cannot be described without the use of such words ['or,' 'not,' etc.]." He implies that the description of the belief must contain the expression of the belief. But the way in which he develops his view is open to criticism. He writes: "If there were no such mental phenomena as doubt or hesitation, the phrase 'today is Tuesday or Wednesday' would be devoid of significance. The non-mental world can theoretically be completely described without the use of such logical words as 'or', 'not', 'all' and 'some'; but certain mental occurrences – e.g., my belief that today is Tuesday or Wednesday – cannot be described without the use of such words. This is one important respect in which, on my view, logic is dependent upon psychology." This whole train of thought is surely erroneous. If the word 'doubt' is used to express the fact that a person is in a mood of uncertainty or that an animal behaves in a characteristic way, this feeling or this behaviour can be completely described without the use of such logical words as 'or' and 'not'. But what Mr. Russell has in mind are, of course, other cases of doubting. He appears not to notice that, when he is describing a doubt as to whether today is Tuesady or Wednesday, he is describing an expression of the doubt and, therefore, has to use logical words. Mr. Russell is misled by the ambiguity of such words as 'mental phenomenon', 'mental occurrence', which are used in such a way that they designate in some cases what may be called private experiences, in other cases operations with symbols or dispositions to do such operations.

I agree with Mr. Russell that our logic is not a system unalterable to all eternity like the Platonic ideas, but that the way in which logic is embodied in our ordinary language and the direction in which we develop it when we construct a formal system is induced and guided by various occasions and conditions of our life. In this connection, I should like to say that there is a wide field of most fascinating questions as to the relation between logic and life. But it seems to me that the example chosen by Mr. Russell, with its appeal to such phenomena as belief and doubt, is unfortunate.

I have said that the rules of our logic are induced by certain experiences, so that if these experiences had been different, another system of rules would be more suitable, in the same way in which one system of geometry can fit a given system of physics better than another one. This raises questions of a peculiar and difficult kind. Let us imagine a language which contains merely the words 'and' and 'if', but not 'or'. What would be involved in saying that there are certain circumstances by which the members of such a tribe would be induced to invent the concept 'or'? What would such a process be like? Shall we say that the introduction of the concept 'or' would be caused by these circumstances? What would it mean to speak of causation in such a case? I will not deny that one can look at things from such an angle. But I would prefer to compare this case with the discovery of a new system in logic or mathematics. Should we say that a mathematical discovery is 'caused' by the circumstances which preceded it? Certain situations may be the occasion for people to invent a concept like 'or'; nevertheless, there is a gulf between these situations and the invention of the concept 'or'.

Therefore, the problem lies deeper than Mr. Russell seems to think. In any case, it would be an over-simplification to say that words like 'or' are read off from a state of mind, e.g., from a feeling of uncertainty, as a tune is read off from the notes. Such a view would imply that the idea of 'or' was pre-existent in the mind, and that all we have to do is to translate it from an amorphous form into the articulate form of language.

III

What puzzles Mr. Russell most is the relation between an object-word and its meaning. He says that if I describe what I am seeing at this moment and utter certain words, this utterance is causally produced by the situation; and that the study of such causal processes is necessary for getting a clear understanding of the function of our language. I think that this view is the result of a misunderstanding, upon which I hope to be able to throw some light.

Let us start with the idea of a transition, which may be explained by a few examples:

(1) A move in the game of chess is the transition from one position of the chessmen to another according to the rules of the game.

(2) Suppose someone is addressing envelopes and using a list which correlates the names of people with their addresses. In looking up the address of a person he is making a transition in this list.

(3) Another example would be the use of a list of colours. If the name of a colour is called, I have to pass over from the name to the corresponding colour and to copy it.

(4) In solving an algebraic equation, I have to transform it step by step.

Each of these steps is a transition according to the rules of algebra.

(5) Every deduction is a transition from one proposition to another in a calculus.

What then is the difference between a causal connection and such a transition? For example, what is the difference between a calculation made by a machine and that made by a person? An important difference is that a person's calculation can be justified by rules which the person can give when he is asked for; not so the calculating of a machine, where the question, 'Why do these numbers appear?' can only be answered by describing the mechanism, that is to say, by describing causal connections. On the other hand, if we ask the person why he has calculated in such a way, he will appeal to the rules of arithmetic. He will not reply by describing the mode of action of a hidden mechanism in his brain.

Does this mean that we contrast the act of calculating with the working of a machine? Or, to put it more exactly, that the behaviour of the calculator does not obey causal laws? By no means. We do not deny that his behaviour is caused by the situation and by previous circumstances. He would not act as he does if he had not undergone a process of education. But this process of education, as Mr. Russell admits, is irrelevant.

Let us imagine a case in which a person has to call out the names of the colours which a lamp shows. There are two ways to consider it:

(1) The colour of the lamp automatically causes the uttering of the word.

(2) The case is a transition from the colour of the lamp to the word according to given rules.

From this example we can see the two ways in which we can look at language. Language may be regarded as a kind of mechanism, to explain the mode of operation of which is a task for psychology. But this is not the way in which a logician describes language. What he is interested in is the geometry of language, not its physics.

What we have said can be expressed by saying: We look upon language not as a mechanism, but as a calculus. To put it more accurately: We *compare* language to a calculus. It would not be correct to say: language is *not* a mechanism, it *is* a calculus. There is no question that words produce many various effects and are in their turn caused by various processes. All we want to maintain is that the logician considers language, not as a mechanism, but as a calculus. In saying this, we are not making any statement about language, but are giving the point of view from which the logician wishes to consider language.

The calculus proceeds no matter what are the causes which determine its separate steps. If a person paints a surface red at the command 'Red!' this process may be regarded as a transition in a calculus. The actual procedure may be:

(1) The word 'red' has been explained to him by a demonstrative definition. Before carrying out the command, he recalls the colour of the specimen.

(2) The word produces his action in an automatic way.

(3) The person uses a list correlating colour-words with colours. If he hears the command 'Red!' he looks for the corresponding word on the list, passes across from it to the colour and copies it.

The fact that he obeys the command, i.e., that he chooses the right colour, may be causally explained by the hypothesis that there is a linking mechanism of a particular kind; but whatever the mechanism, one and the same transition is represented.

The explanation of a word can play a double rôle: –

(1) The explaining is the cause of the word's being used in a particular way;

(2) The explanation is the ground for this use.

We should not use the words of our mother-tongue in the way we do, unless we had learned this use. In this sense a word's having been explained to me is the cause of my understanding of the word; but this account would not justify the use of the word. There is a connection between the two rôles of explanation, and it is this fact which misleads us into thinking that the meaning of a word is the way in which it functions causally. It would be more correct to say that the meaning of a word is its purpose. For purpose and causal functioning are connected in such a way that an effect which never occurs would not be said to be purposed.

If we explain a word by means of a demonstrative definition, we use a gesture to guide the eyes of the other person in a certain direction. Influences such as these play an important part in our original learning of language. In speaking of training, we lay stress upon the causal, in using the word 'explanation', the normative aspect; in the latter case we *compare* the words and the gestures of a demonstrative definition with rules in the fully-developed language.

CHAPTER VII

NOTES

[1] I wish to emphasize my indebtedness to Dr. Wittgenstein, to whom I owe not only a great part of the views expressed in this paper but also my whole method of dealing with philosophical questions. Although I hope that the views expressed here are in agreement with those of Dr. Wittgenstein, I do not wish to ascribe to him any responsibility for them.

[2] [M. Schlick, 'Über das Fundament der Erkenntnis', Erkenntnis 4 (1934) and frequently thereafter.]

[3] [Bertrand Russell and R. B. Braithwaite were the other two contributors to the symposium from which this paper is taken.]

CHAPTER VIII

WHAT IS LOGICAL ANALYSIS?*

It has often been noted that philosophy and science express two very different types of attitude of the human mind. The scientific mind searches for knowledge, i.e., for propositions which are true, which agree with reality. On a high level, it rises to the construction of a theory which connects the scattered and in their isolation unintelligible facts and in this way explains them. But the philosopher cannot be satisfied with this. The very nature of knowledge and truth becomes problematic to him; he would like to get down to the deeper meaning of what the scientist does. Now what can be gained through philosophy is an increase in inner clarity. The results of philosophical reflection are not propositions but the clarification of propositions. Wherever real progress has been made in the history of philosophy, it resided not so much in the results as in the attitude to the questions: in what was regarded as a problem, or alternatively, in what was recognized as a falsely formulated question and excluded as such. Thus when Hume showed in his famous critique of the concept of causality that we only perceive the succession of events and never an inner bond that ties them together, the permanent gain from his reflection did *not* reside in a philosophical proposition – an axiom around which other propositions cluster as around a crystal of truth – but in the clarification of the *sense* of causal propositions; and hence not in an increase in the number of propositions but rather in its diminution: in the disposal of all that baggage of seeming truths and imagined knowledge that trailed behind that false idea. Hume analysed the concept of causality; and in this sense, philosophy can be called the logical analysis of our thoughts.

But what does this logical analysis consist in? It seems to me that the correct philosophical attitude depends to a large extent on clarity at this point, and it is therefore in our interest to become more familiar with this concept.

Analysis means dissection, dismemberment. Thus logical analysis seems to mean: dissection of a thought into its ultimate logical elements. And here

* First published in German in *Erkenntnis* 8 (1939–40), 265–289. Reprinted in Reitzig.

we are only too ready to call to mind analogies from various other fields: As the physicist analyses white light through a prism and breaks it up into the various colours of the spectrum, as the chemist analyses a substance and uncovers its chemical structure, so, roughly, do we imagine the business of the philosopher: his job is to lay bare the structure of a thought, its logical construction. Is this comparison correct?

The question known as the problem of *elementary propositions* seems to me to provide a good example for this purpose. This question is connected with certain more profound investigations into the structure of our language, which will now be sketched in a few words.

1

Propositions can be tied together and assembled into higher units, and this can be done in various ways. Language employs different conjunctions for this purpose, like 'and', 'or', 'if', 'because', 'although', 'after', etc. These can be put into two sharply separated categories, which will be illustrated by the following examples:

> 'it has been getting warm and the sun is shining'
> 'it has been getting warm because the sun is shining'

The characteristic difference between the two kinds of conjunctions is this: In the first case I only need to establish the truth of the individual propositions to be certain of the truth of the whole; but not in the second case: even if it has been getting warm and the sun is shining, there need be no causal connection between the two processes. The truth of the complex proposition is not yet guaranteed by the truth of its parts.

A complex of the first kind is called a '*truth-function*'. More precisely: p is called a truth-function of the propositions p, q, r (its arguments), if the truth or falsity of p depends *only* on the truth-values of p, q, r ... The simplest examples are 'p and q', 'p or q', 'if p then q'. If we consider that an unlimited number of combinations can be formed from n given propositions by means of 'and', 'if', 'not', etc., we are easily led to the opinion that there is an infinite number of such strings. In fact, their number is well limited, for it turns out that there occur repetitions in the series of strings, i.e., combinations which look quite different outwardly but which express exactly the same sense. The best way to see this is to make use of a different representation which brings the structure of such a complex into full view.

It is, evidently, essential to a truth-function that it be true for certain truth-values of its arguments and false for others. A truth-function is therefore completely determined by giving the cases in which it is true and the ones in which it is false. For two propositions which are independent (i.e., which neither say the same nor contradict each other, and of which neither is a logical consequence of the other), there are four different possibilities of truth or falsehood, which can be arranged in the form of a table:

p	q
T	T
T	F
F	T
F	F

We get a truth-function if we write down for each of these possibilities whether the whole proposition is true or false. Thus

p	q	
T	T	F
T	F	F
F	T	F
F	F	T

represents the proposition 'neither p nor q'; for we have excluded all possibilities except one: that p and q are both false. The proposition 'if p then q' would be expressed in this notation in this way:

p	q	
T	T	T
T	F	F
F	T	T
F	F	T

There will be altogether sixteen such truth-functions if we distribute the signs 'T' and 'F' in all conceivable ways over the schema of the four truth-possibilities. Out of n propositions we can form 2^{2^n} truth-functions.

It is easy to see that the truth-functions really capture what we mean when we utter a proposition. To put it briefly, let us say that a simple proposition communicates a fact; though a combination of propositions will not then communicate a combination of facts. In saying 'if it is raining, it is wet', I do not assert the existence of a certain state of affairs; what I want to say is: 'It may be that it is raining and that it is wet, or that it is not raining and that it is wet, or that it is neither raining nor is it wet'. I do not therefore use the proposition to specify a fact but a scope or range of facts; and this is precisely what the truth-function expresses. Reality is not yet fixed by the proposition; the proposition picks out a group of possible combinations of states of affairs and asserts that one of these possibilities is realized in the world.

The truth-functional representation makes the logical structure of thought stand out more clearly than ordinary language. Thus if we compare the three propositions, 'it is not raining or it is wet', 'if it is raining, it is wet', 'if it is not wet, it is not raining', we cannot tell at first glance that they express exactly the same thought. The truth-functional representation first uncovers the common internal structure of these propositions.

Among the truth-functions that can be formed out of n propositions, there are two extreme cases: one is that the proposition expresses agreement with all truth-possibilities, i.e., that it is *always* true; the other is that it is *never* true. In the first case the compound proposition is called a tautology, in the second a contradiction. Tautology and contradiction are *limiting cases*, degenerate forms of a compound proposition: their truth or falsity no longer depends on the behaviour of the real world, and they therefore do not tell us anything about the real world either.

It may be worth while to bring up a point which is not always clearly understood: If tautologies do not say anything at all – why is it, then, that they play such an important part in logic? What is the use of setting up a system of propositions without any content? The answer is that the logician uses a tautology as a *means* for demonstrating purely logical relations between propositions. Thus if the proposition q folllows from p, this can be seen by noting that the compound proposition 'if p then q' is a tautology. A contradiction could, of course, serve the name purpose: it would be equally possible in principle to formulate the rules of inference by means of contradictions. But it should be emphasized that – in the elementary part of the calculus – tautology and contradiction are entirely dispensable, since all logical relations can be recognized by the mere look of the propositions in the corresponding notation. If two propositions are written as truth-functions, we can easily read off from their structure whether the one follows from the other; whether they contradict each other, etc.; the former is the case when all the truth-grounds of the one proposition (i.e., all the truth-possibilities it affirms) are also truth-grounds of the other, and hence, when the scope of the one is wholly contained in the scope of the other.

The idea of a truth-function can be extended to cases in which the arguments are not given individually but in another way. The propositions 'this leaf is green' and 'this paper is green' are both of the form 'x is green'. After classifying propositions by their form (irrespective of their truth or falsehood) and subsuming them under a symbol, a 'propositional function', we can, in thought, form their conjunction (their logical product) or their disjunction (their logical sum). The product of all propositions of the form 'fx' is expressed by writing '$(x) . fx$', and their sum by writing '$(\exists x) . fx$'. '$(x) . x$ is green' would then mean that all things are green, and '$(\exists x) . x$ is green' that there is at least one green object. Accordingly, propositions containing 'all' and 'there is' also prove to be truth-functions, although their arguments are not mentioned explicitly but only characterized by their form. The reason why they can be called truth-functions is that their truth or falsity depends only on the truth or falsity of the propositions falling under them.

II

So far we have described how, starting from given propositions, we can ascend to complex propositions, and we have introduced the common principle in the formation of these structures. But this relates only to the first category of constructions. This brings us to the question: What about constructions of the form 'it has been getting warm *because* the sun is shining'? It is clear that these are not truth-functions of their component propositions. But this is not yet to say that they cannot be resolved into truth-functions of other propositions. On the contrary, we are easily led to attempt such a resolution. Since the proposition asserts a causal connection, and since the characteristic mark of such a connection is the regularity with which cause is followed by effect, we are easily led to analyse the proposition as follows: 'The sun is shining; and whenever the sun is shining, it gets warmer'. The latter proposition is nothing but a summary of all particular propositions of the form 'If the sun is shining at time x in place y, then it gets warmer at time x in place y'. We should thus have analysed the proposition in the end into nothing but particular propositions. That we cannot really write these down appears on this view as a human weakness, which in no way prejudices the possibility which exists at least in principle. (Such a view was formulated by F. P. Ramsey in *The Foundations of Mathematics*.)

This brings us face to face with a very general problem whose clarification will occupy us in what follows. The problem is to decide whether all complex propositions can be resolved into truth-functions. Compound propositions containing words like 'although' and 'even if' will here be excluded, since they evidently do not report objective facts but express the attitude of the speaker. As for the remaining complex propositions, in so far as they describe facts in the real world, we seem justified in supposing that they admit of such a resolution; and at the same time this representation seems to be the way to penetrate more deeply into their logical structure.

III

If we try to analyse the propositions of our language, we run into a major difficulty: a proposition can look simple without being simple. Anyone who has concerned himself with the logic of our language, no matter how superficially, knows how little weight is to be placed on the outward linguistic form. Take a simple example: The proposition 'the electrical field-strength here is such-and-such' appears simple; but if we go back to the definition of the concept 'field-strength', we find that the proposition says: 'If a test-charge of such-and-such a force and such-and-such inertia is placed at this point in space, then it undergoes such-and-such an acceleration'. This conditional proposition takes the place of the apparently simple statement about field-strength. The concepts of electric charge, inertia, and acceleration appear as the given elements in the new statement. These in turn stand in need of definition, and the next step would accordingly be the elimination of the concept of charge, the following one the elimination of the concept of inertia, etc., till we came in the end to a combination of statements which dealt only with concrete sense experiences, like a pointer indicating a certain point on a scale. Now it has been thought that such a system of relations expressed exactly the same state of affairs as the statement about field-strength and could well be substituted for it. If we compare the two statements with each other, we are at first unable to see why the latter should be preferable to the former. We could say instead that it was much more complicated, for the structure of the statement has become more complicated at

each step in the transformation. But *one* circumstance gives the latter statement a tremendous advantage: there is no longer any reference to imper-
ceptible entities like 'field-strength', 'charge', etc.; instead, this statement
describes a system of relations which refer exclusively to facts of observation. From this it has been inferred: If a statement of physics or some other
science mentions things which elude direct perception, then the statement
can always be formulated in such a way that it only contains relations between observables. This only means that any statement must be capable of
being verified by observation, and this is obvious, for otherwise it would
have no sense. Abstract concepts like 'field-strength', 'charge', etc. could
thus be banished altogether from the scientific vocabulary, without sacrifi-
cing anything of the content of scientific propositions. Such a reconstruction would be possible, even though infinitely complicated. The mathematician will be familiar with a very similar situation: Any proposition of analysis, no matter how remote it seems to be, must be capable of being expressed in a form in which it mentions only natural numbers. Thus the
insight that π is transcendent can be retranslated into certain relations between natural numbers. Yet hardly anyone has even made the attempt to
represent the proposition in this way: the statement would soon become
too complicated and too difficult to survey.

While the carrying-out of such an analysis runs into various practical difficulties, this only increases the theoretical interest of its possibility. In the
case of mathematics, it first brought the nature of rational numbers into
full view: it led to the recognition that any statement about irrational numbers can be formulated in such a way that it deals only with series or sets of
rational numbers; the rational numbers go back in turn to integers and
these, finally, to natural numbers. The result is a hierarchy of mathematical
concepts. If we consider the possibility of reduction in the case of an empirical theory, its concepts will also be arranged as on a genealogical tree where
each higher concept is 'constituted' by certain lower ones. (Such considerations were pursued by Carnap in *The Logical Structure of the World*.)

We thus get the following picture of the structure of the realm of concepts: Suspended from the highest abstract concepts of a scientific theory,
there is an uninterrupted chain of interlocking definitions which establish
a connection with the most concrete things of our experience – or if you
prefer, with the words which describe our experience. As we transform a
proposition of the theory by gradually going back along the chain, the

proposition resolves itself into an increasingly complicated complex of propositions which become more concrete and come closer to reality with each step. When the scientist reaches the concepts of everyday life, he rests content and asks no further questions. And this is right, for he thinks he understands the sense of these propositions well enough. But are we to conclude from this that no further analysis is possible? Or that we have now advanced to the ultimate elements of representation? Certainly not. The concept of a pointer moving along a scale is from the point of view of daily life simple and unproblematic; but what are we to say from the logical point of view? Does it really admit of no further resolution? As the expert knows, this question touches on what is called the problem of substance in classical philosophy. It has often been said that statements about bodies go back ultimately to statements about perceptions, and that bodies do not therefore represent the final (irreducible) element in the structure of reality. Such efforts to resolve the concept of a substance go back as far as Locke and Berkeley, where they led roughly to the view that a body is a relatively constant complex of ideas, and they have been carried further in our time with the methods of symbolic logic, without, however, yielding a single solution (cf. the investigations of Russell, Whitehead, and Carnap). Progress has, however, been made in *one* respect: Today we would no longer try to resolve a body into a complex of sensations; we would rather formulate the core of the problem in this way: Can statements about bodies be transformed in such a way that they deal only with perceptions? More precisely: Are the statements of the former kind truth-functions of statements of the latter kind? Thus what is hidden behind those centuries-old lines of thought is, again, the problem of the structure of our propositions. The state of the question shows at any rate that we must face the possibility of further analysis. In our search for the ultimate elements, we may have to descend below the level of the concepts of daily life, by resolving the concept of a body also in the way suggested. And if we do this, we will only be continuing the chain of concept formation in the opposite direction.

Where will we finally get to if we take this road? In the case of mathematics, the chain of concept formation comes to an end as soon as we reach the level of natural numbers. No further reduction is possible, at least within the framework of present-day mathematics (though one modern school of logicians tries to push the analysis still further). Where does the analysis of empirical propositions lead to? Or is there no end at all here? Can analysis proceed to infinity?

IV

One answer to this question was given by Wittgenstein: according to him, analysis must eventually lead to ultimate elements, to absolutely simple propositions. Wittgenstein calls these formations *elementary propositions.* They are, as it were, the atoms out of which the universe of propositions is built. If this view is correct, we would get an astonishingly simple picture of the structure of language. According to it, all propositions of our language would be laid out according to a simple and perspicuous plan: they would all be truth-functions of elementary propositions. If we were acquainted with the totality of elementary propositions, we could derive from it in theory any proposition, no matter how complicated. A statement about reality would mean that reality agreed with certain combinations of elementary propositions but not with others. How can such a claim be justified?

Two things would be needed: First, a demonstration that a proposition can only be dissected by analysis into truth-functions of other propositions, and hence, that the truth-functional schema is the only principle of the formation of propositions. Secondly, a demonstration that the analysis must come to an end. We shall deal here with the second point first and reserve the right to return to the first after we have cleared up the second. We therefore assume provisionally that it has already been shown that a proposition can only be resolved by analysis into truth-functions of other propositions. The following line of thought then compels us to assume that there are elementary propositions: Since the truth or falsehood of a truth-function depends simply and solely on the truth-values of its arguments, such a proposition can never be directly compared with reality, but only indirectly, via its component parts. Thus when I say 'if p then q', I can indeed compare the individual propositions p and q with reality and thus establish the truth of the whole proposition, but not the compound proposition itself, for nothing in reality corresponds to the word 'if'. Now if the analysis of the proposition led to infinity, i.e., if the proposition could be dissected into component propositions and these in turn into other propositions and so on without end, then there would be no possibility of testing the proposition against reality. In the attempt to verify it, we would again and again be led to propositions; thus the truth of the proposition would always depend on the truth of other propositions and never on the facts. If we take the view that the sense of a proposition lies in the procedure by which it is to be veri-

fied – and verified by comparison with the facts – then we would have to say
that the proposition cannot be understood at all. But we do understand the
propositions of our everyday language, and this seems to yield the conclusion that analysis must reach an end.

On the basis of this line of thought, Wittgenstein assumed the existence
of elementary propositions. What do we know about them? First, that they
are logically simple, i.e., that they are not truth-functions of other propositions. This means that they must be immediate combinations of primitive
signs, i.e., of signs which cannot be dissected by any further definition. Now
if p and q are two elementary propositions, then the combination 'if p then
q' can certainly not be a tautology, and this seems to yield the further conclusion that one elementary proposition cannot be inferred from another,
that elementary propositions are independent. Finally, it is part of the concept of an elementary proposition that it describes reality completely, that
it leaves nothing indeterminate. If a proposition p were to leave some scope
to the facts, we could imagine this scope divided into two parts, with two
proposition p_1 and p_2 corresponding to them; the proposition p would then
evidently say the same as 'p_1 or p_2'. Thus if a proposition leaves a certain indeterminacy to the facts, this is an indication that our analysis has not yet
come to an end. An elementary proposition seems, then, to be characterized by the following properties:

(1) it is simple;

(2) elementary propositions are independent of each other;

(3) an elementary proposition leaves reality no scope.

V

Are there propositions which satisfy these requirements? Can we give an
example of an elementary proposition in our language? As everyone
knows, the sense of a proposition can often be explained in other words,
viz., when the proposition contains signs or words which can be explained
by means of other signs (cf. the example of field-strength). But in the end we
arrive at propositions where we are at a loss for an explanation; when we
come to 'the pointer of the apparatus indicates point 5' or 'this is blue', the
sense of our statements seems to be so clear and definite that it is no longer
possible to paraphrase them. Whatever we wanted to say would be less
clear than the given statements. Shall we perhaps regard them as the ele-

mentary propositions we have been looking for? The two following reasons seem to tell against this:

First, such propositions are not independent of one another; thus if I say in looking at an area 'it is blue', then it follows from this that it is not red; i.e., the conjunction of the two propositions, 'this area is blue and it is not red', would imply a contradiction. If these propositions were elementary, this would be incomprehensible. That they yield a contradiction indicates that these propositions already possess a certain logical structure and that they are not truly elementary. What kind of structure this is, is indicated by the following consideration: If proposition q follows from proposition p, then p must already be contained in q – like a factor. By way of example, the proposition "I am busy Monday night" follows from the proposition "I am busy every night this week", and here the latter proposition can in fact be analysed in such a way that the former occurs in it. We shall be tempted to imagine the relation of the two colour ascriptions above in exactly the same way: If we only had sufficient insight into their true logical structure, we should be able to read off from it that the second proposition was contained in the first.

Secondly, while such a proposition describes an observable state of affairs, it can never describe it accurately: no matter how carefully I express myself, there will always be a certain vagueness attaching to my statements. When I say 'the pointer indicates point 5', I have neglected to give the exact form of the pointer, and further, the colour of the background and various other circumstances; and even if I wanted to complete my description, I could not possibly reproduce the hue, the brightness, or the contours with absolute precision. Language disposes only of a limited supply of signs; its palette is too poor to reproduce the wealth of appearances; and it can therefore only perform its task by simplifying, schematizing the facts. Each proposition only prepares the frame as it were, it constructs a conceptual scaffold within which reality is free to vary. Our language proceeds like a draughtsman who only makes a sketch of an object. In short, each of these propositions gives reality some scope and seems therefore capable of still further resolution.

But now we get into a new difficulty. For the reasons given, it seems to us that these propositions cannot yet be the end-points of analysis; and yet we are at a loss when asked to specify what a real elementary proposition looks like.

But is the reason for our entanglement not obvious? Our language is only equipped for the needs of our lives; it is too coarse to reproduce the ultimate, that which matters to us. Could we not devise a language with a finer structure capable of performing this task? This would, of course, entail a change in the aim of our investigation. We would no longer be asking what elementary propositions looked like in *our* language, but what forms we would arrive at if we imagined a continuation of the process of analysis beyond the store of propositions of our actual language.

VI

When we try to create a symbolism with which to capture our immediate experience – the experience 'that slips away from us' – in signs, we must first attend to the structure of the phenomena themselves and observe how they can be reproduced in symbols. And here we must squarely face the possibility that the customary means of expression of our language – such as subject-predicate propositions, two-place relations, etc.–may break down, and that we may have to construct a symbolism of a very different logical form. Let me explain this by an example. If I want to describe my room – I now mean the visual image I see – I shall describe, e.g., the position of the furniture in the room, i.e., the spatial arrangement of objects. But I can also give a very different form to the description, such as this: I represent the surface of the room, or that part of it which I see, by an equation of analytical geometry and give the distribution of the colours on this surface – in this colour field. In this form of description there is no longer any talk of individual objects like tables, books, and chairs and of their spatial position; this symbolism recognizes neither subject nor predicate, nor relations – all this appears only when we describe the facts with a schema of a quite particular kind: with the grammatical forms of our ordinary language.

In the search for elementary propositions, we may therefore have to reach out for entirely new means of description – for structures which can be used to reproduce adequately the perceptual field and the changing phenomena within it. In pursuing this path, we might arrive at important insights. May not certain difficulties in the reproduction of immediate experience which have been pointed out by Bergson and the phenomenologists – such as the duration of the present – be somehow due to the fact that we employ a language whose structure does not quite fit the phenomena it is supposed to describe? And could not this difficulty be overcome by using a language with the right multiplicity? A new perspective thus seems to open up before us: Behind the propositions of our ordinary language stand other formations whose structure we can today discern only indistinctly. To make them clear and explicit appears to be an important aim of epistemology.

All these considerations stand or fall with the assumption that it is really possible to resolve a proposition into ultimate elements. A simple example should show whether this assumption is correct. The position of a patch in the visual field can be described in these words: It lies somewhere in the square $ABCD$. A certain indeterminacy attaches to this description, as indicated by the word 'somewhere'. According to what was said above, we are to imagine that this proposition can be further analysed. But what is this analysis supposed to look like? Is it like this: the patch is in this → place or in this → place ... or in that → place? But if we were asked how many possible positions there are for the patch in the square, we should be at a loss for an answer, although we might perhaps be able to draw certain positions. Is this because there are infinitely many positions here in the strict sense of the word? To get clear about this, let us first consider a simple example and ask: How many colours do I see when I look at a spectrum? Shall we say: infinitely many? And shall we perhaps define the kind of infinity in question by making use of the concepts of set theory? This would be straying into the wrong system of thought. The colour continuum has, in fact, a very different structure from the number continuum. Whether two real numbers are the same or different can be settled unequivocally. No matter how close they come to each other in the number series, they still remain different. A colour, however, merges into the next one; or more correctly: it makes no sense to speak as if the colour continuum consisted of individual atoms, i.e., out of sharply defined hues. On the contrary, this continuum has a peculiar indistinctness, which excludes the application of the concept of number. Only what can be clearly distinguished can be counted. 'I see infinitely many colours' only means that it makes no sense to say: I see only twenty colours or I see only thirty colours or I see only forty colours. So when we say that we see *countless* colours, our language expresses a perfectly correct feeling: it forbids us to give the number of the colours, and that is all it does.

The question how many positions there are for the patch in the square can be given a very similar answer. I could, of course, try to enumerate the positions by saying: The patch may lie here → or here → or here → and so on; and here the word 'and so on' seems to point to infinity. Our only comment on this is that the word 'and so on' has at any rate a different grammar in this context than in the case of the series

1, 2, 3, 4, 5, and so on,

where it points to the formation rule of the series. In the latter case I can say with emphasis: 1, 2, 3, 4, 5, and *so* on; but not in the example of the positions of the patch, for here I would be asked at once: 'And *how* on?' 'There are infinitely many positions of the patch in the square' means, again, no more than this: it is nonsense to say that there are so-and-so many different positions. This has nothing to do with infinity in the mathematical sense.

The proposition 'the patch lies somewhere in this square' cannot therefore be represented either as a finite or as an infinite disjunction; and this means that it is *not* a disjunction, although it is in another respect similar to one. It can thus be shown, by reflection on even the simplest example, that a proposition cannot successfully be resolved into ultimate elements, and we are slowly beginning to see that in dealing with this question we have been moving in the wrong direction.

VII

Wittgenstein was the first to recognize the way out of these difficulties, and we shall follow him in our account. It is quite true that two logically simple propositions cannot contradict each other in the sense that their logical product yields a contradiction; but they can exclude each other, and this is true even of many propositions of our language. Thus the propositions 'this stick is 2 metres long' and 'it is 3 metres long', 'Mr. N. is 30' and 'he is 40 years old', 'At this point it is now 10 degrees' and 'it is 15 degrees' are incompatible, and not because experience teaches that such states of affairs never coexist, but because these combinations of propositions mean nothing – because they are senseless. That is to say, it is part of the meaning of the expressions 'length', 'age', 'temperature', etc. that a staff has only one length, a man only one age, and a point in space only one temperature, and when we ascribe two lengths to a line, we offend against the rules of logical grammar and talk nonsense.

How is this rule expressed in our logical symbolism? The conjunction of two propositions is expressed by the schema

p	q	
T	T	T
T	F	F
F	T	F
F	F	F

This presupposes that the two propositions are *independent*, i.e., that each of them can be true or false irrespective of the other. But if we are dealing, e.g., with the two propositions 'this staff is 2 metres long' and 'this staff is 3 metres long', this notation leads us astray by placing more possibilities at our disposal than correspond to the facts, i.e., than we can meaningfully admit. We must therefore correct the notation by excluding one of these moves: the one which allows the truth of p and q; accordingly, we strike out the first line of our schema; and the result is that a logical product – or rather, what corresponds to a logical product in the corrected notation – is represented by the following schema:

p	q	
T	F	F
F	T	F
F	F	F

But this is the expression of a contradiction. It can be shown in exactly the same way that the proposition 'this point is not red' (= not q) follows from the proposition 'this point is blue' (= p). This amounts to the proof that the compound proposition 'if p then not q' is a tautology. If we take the representation of this proposition:

p	not q	
T	T	T
T	F	F
F	T	T
F	F	T

and strike out the second line, which expresses the combination 'p is true', we get a tautology. Thus it is not true at all that we must wait for the logical analysis of these propositions before we can understand their structure; the truth is that this structure comes out in the actual use of these propositions; but we already know this use, and we only need to set it out in clear rules.

These considerations throw light on another question which has been responsible for much misunderstanding: the question to what extent the consequences of a proposition are thought along with the proposition. If I
know that a bluebell is blue, then I also know that it is not red. We are now
inclined to say: the latter proposition is already contained in the sense of the
former; it only needs to be taken out of it as it were. But how is this to be understood? Does it mean that whoever utters the former proposition thinks
silently of the latter proposition? This would not correspond to the truth,
for if we asked someone what he was thinking of when he uttered the former proposition, he might well reply: I was thinking of what I was saying.
We are now tempted to say that this proposition was thought along 'sub-
consciously'. And now the whole affair appears in a somewhat mysterious
light. The mystery can be cleared up by noting that the rule of inference 'q
follows from p' can be written in the form of the equation

$$p \cdot q = p.$$

This way of writing it shows that the conclusion of an inference is in fact
contained in the premise, viz. in the sense that the conclusion, added to the
premise, yields the premise. The proposition 'a is not red' is in fact contained in the proposition 'a is blue', but, it should be noted, only in the sense
that there exists the rule

a is blue and a is not red $= a$ is blue.

What is right about the view that the one is thought along with the other is
only what is expressed in this rule of inference; to the extent that this rule
holds, we can say that the one proposition is contained in the sense of the
other. But this is not, of course, to make a psychological statement but to
give a grammatical connection.

The second reason why we did not want to recognize such propositions
as elementary propositions was that they gave reality some scope, just like
truth-functions; and so it seemed to us that they had to be capable of further resolution. This was a fallacious inference, produced by the ambiguity
of the expression 'scope'. If I say in the case of a truth-function that the
proposition leaves reality some scope, I can state with precision what this
scope consists of: it consists of those individual truth-possibilities of the
table for which the proposition is true, and these form a clearly defined system. By widening this scope step by step, I can arrive at a tautology within

the notation in a finite number of steps, and by narrowing the scope, at a contradiction. But if I say that a proposition of the form 'this is blue' leaves reality some scope, this only means that the proposition is vague, that there is a certain indeterminacy attaching to it – an indeterminacy which, it should be noted, can no longer be resolved into individual clear-cut cases. While it is possible in the former case to give the distance of a proposition from a tautology, this is impossible in the latter case. In short, the word 'scope' has a different meaning in the two cases, even though there is a certain similarity between the two ways of using it; and this is precisely what misled us into thinking that the two kinds of formation had the same underlying structure.

Now what about the logical structure of our language? Is it an atomic structure? If by an elementary proposition we mean a proposition which cannot be analysed into a disjunction or conjunction of other propositions, then we can safely say: Yes, there are elementary propositions, and many propositions of our language are among them. There is now no longer anything mysterious about this concept. We need no longer assume that hidden behind the propositions of our language is a world of different formations whose structure we can today discern only indistinctly, as through a fog.

VIII

Only a few words will be devoted to the further question: what is the position of hypotheses in our system of propositions? Is a hypothesis only a summary of propositions about past and future observations? I do not think so. The true relation between hypotheses and particular propositions is much more complicated and can only be hinted at within the framework of this little sketch. This connection has been imagined up to now in too simple a way, when it was thought that hypotheses are general propositions by means of which we *infer* future observations from known data. If there were really a strict inferential relation here, then a hypothesis would be refuted by a single case which failed to fit it; but there is no question of this in actual scientific practice. A law of nature which has stood the test till now, which may even have become the corner-stone of an important theory, will not so easily be shaken by individual observations. Which astronomer would give up Kepler's laws on the basis of *one* observation? If such an observation were actually made, we would first try out various other possible explanations (deception of the observer, deviation of the planet due to unknown heavy masses, other kinds of disturbances, e.g., due to friction with thin gases, etc. etc.) and only if the resulting structure of hypotheses became too complicated, if it no longer satisfied our craving for simplicity and claity, would we decide to drop those laws. And even then the 'refutation' would not be definitive, valid for all times; for it could always turn out that some circumstance had eluded us which made the whole appear in a different light when it was taken into account. The history of science records a number of cases where the apparent defeat of a theory was turned into a complete victory (Olaf Römer, Leverrier). Since we cannot foresee the direction of scientific research, we shall always have to face – in principle – the possibility of such a rehabilitation.

We had therefore better formulate the relation more carefully, by saying that certain observations tell for or against a hypothesis, which does not mean that they confirm or refute it. How much value we attach to a contrary observation, and when we regard it as a 'refutation' of a theory, depends on the whole scientific situation, and it appears to be a hopeless task to set up precise rules here.

From all this it appears that the connection between hypothesis and observation is looser than has been imagined up to now by logical theorists. We could, incidentally, arrange the propositions of our language in certain strata, by admitting into the same stratum all those propositions between which there exist precisely formulable logical relations. Thus the laws of thermodynamics form a system of statements (equations) whose elements are connected by strict logical relations and within which it can be decided with precision whether two equations contradict each other, whether the one follows from the other, etc. The statements made by an experimental physicist in describing certain observed data, like the position of a pointer in his apparatus, are also related to one another in precisely formulable ways. (If a pointer indicates point 3 on a scale, it cannot possibly indicate point 5; there exists here a strict relation of exclusion.)

On the other hand, a proposition of theoretical physics can never come into strict logical conflict with an observation sentence, and this means that there are no precisely formulable logical relations between the two kinds of propositions. All the relations and connections which the logical calculus places at our disposal hold only as long as we move within a given stratum of propositions. But the real problem begins where two strata as it were border on each other, and it is the problems of this plane of cleavage which today merit the attention of logicians.

It could indeed be said: From the validity of the laws of nature it follows in all strictness that an experiment must turn out so-and-so, provided that no disturbing circumstances enter in. But this does not seem to me to be of much help. For since we have no prospect of controlling all possible disturbances in an experiment and cannot even state them completely, we can never infer with certainty how an experiment will turn out (even if we regard the laws of nature we have assumed as strictly valid); rather, the fact remains that an observation sentence does not 'follow' in the strict sense of the word from the laws of nature and the data that are known to us. I should like to express myself like this in general terms: No matter how far we advance, our knowledge always remains surrounded by an area of semi-darkness, a zone of indeterminacy, in which future discoveries will very often be made. It makes no difference whether we locate this genuine indeterminacy in the system of *conditions* or directly in the logical relations between propositions, i.e., whether we say: the logical relations hold indeed in all strictness, but we never know all the presuppositions in question, or whether we say: there exists no strictly formulable logical relation between hypothesis and observation sentence.

If these considerations are correct, they lead to very serious doubts about the applicability of our logical formalism to the whole system of physics. The situation, as it presents itself, seems to be this: Within physics, different ranges of propositions are distinguished from one another. The methods of inference of our logic are only applicable to the propositions of *one* such range at a time. Logic – even in the more perfect form which Russell, Whitehead, and others gave it – is still inadequate to the task of retracing the more complicated relations between propositions of different ranges.

It is not our purpose to go more fully into these things. But even from these very few remarks it should be clear that a proposition of theoretical physics cannot in any way be conceived as a truth-function of observation sentences. For this would imply the possibility of *strict* verification or *strict* falsification. And the same situation seems to recur when we follow up the connection between statements about bodies and propositions about perceptions. The structure of a proposition about bodies comes out in its use, and its use is not such as to allow us to replace the proposition 'here is a

table' by a conjunction or disjunction of perceptual statements. It may
well be true that the former proposition is somehow 'founded' on the latter:
that in justifying the proposition 'there is a table' we must refer in the end to
certain perceptions; but in formulating this foundation, we must proceed
again with extreme care. Think only of how difficult and complicated it
may be under certain circumstances to distinguish between reality and illusion, if we really try to formulate the relationship in exact terms. Can we
enumerate and list all possible illusions? Can we say, if none of these possibilities arises, that it has now been proved conclusively that there is a table?
None of the attempts that has been made so far to give an exact analysis of
the concept of a body has reached its goal. The schema of truth-functions,
even if we add the formation of general and existential propositions to it, is
not wide enough to encompass the actual wealth of propositional forms in
our language. Instead, we feel impelled towards the more liberal view that
our language contains *different* types of propositions which cannot be constructed out of one kind of proposition – e.g., observation sentences – by
means of certain logical connectives. And the simple and unanalysable
propositions in each of those strata can, if we like, be called elementary. It
will be seen that this turns the concept of an elementary proposition into a
relative one. And it is obvious that the search for elementary propositions
now loses its fascination and its philosophical interest, since we are no
longer advancing towards ultimate forms that mirror reality.

IX

The view we have sketched – that propositions about bodies cannot be resolved into perceptual propositions – is open to a possible objection. Could
we not invent new logical methods which could make such an analysis possible? Here I should like to make room for a remark to which I attach the
greatest importance. In dealing with such questions, we often talk as if logical analysis could bring to light something which we cannot see yet, like the
hidden structure of a proposition. But here we are straying, as so often, in
the paths of a misleading analogy. Imagine that someone reports the discovery of a new, heavy kind of water, and that he actually produces the substance; we must then simply accept the fact. But if someone analyses the
statement p and tells me: the proposition p really means 'q or r or s' – must I
then also allow the result of the analysis? Not in the least. We shall come

closer to the essence of the matter if we say instead that he is *proposing* that we use that proposition *p* in the way described, and that it all depends now on whether I *accept* his proposal. If I do, I only let it be known that I have always wanted to use the proposition *p* in this way. What, then, do I do when I analyse a proposition? I simply give part of its grammar – of the grammar I have always used for this proposition. I merely call to mind a way of using it, but I make no discovery.

But even this way of putting it is not yet quite right; for what is sometimes in question is not how I have used a proposition but how I *wish* to use it. To return to the example of the spectrum, I could say that I saw a well-defined number of colours while counting the barely noticeable colour differences; I could thus resolve the statement 'I saw all the colours of the spectrum' into a *finite* conjunction. But in that case I have not uncovered the hidden structure of this proposition; I have simply indicated how I propose to use the proposition. And now it all depends on whether someone else agrees with this interpretation, i.e., whether he acknowledges it as a formulation of what he had in mind when he used those words. And in this he is completely free. He could also have said: No, when I spoke of all the colours, I did not mean as many as could be distinguished in such an experiment; I meant the spectrum as a whole, without thinking of its resolution into individual colours. This shows that it depends on how we interpret the proposition, i.e., which grammar we decide on. By citing such an experiment, we first determine how the word 'all' is to be understood in this context.

Let us put it this way: The question whether a proposition of our language can be further analysed presupposes that I have a *method* for looking for such an analysis – and a method acceptable to the other person. If there is no such method, it is no use inquiring whether the proposition may not 'in reality' be composite, even though we do not know it today. Such a question sounds much too absolutist; it makes it look as if the proposition was simple or composite in itself – whereas it depends entirely on what we determine about it. That is, it is a matter not of analysis but of synthesis, namely of the synthesis of the grammar through a detailed statement of the rules. Analysis is involved only to the extent that we have some prior notion of the rules and then analyse that notion.

We can now also understand why the result of logical analysis is not a proposition but the clarification of a proposition. For clarification consists in nothing but the articulation, formulation, and conscious realization of the rules which are nowhere expressly formulated but which are presupposed in our understanding of everyday language. When Hume analysed the concept of causation, he only set out the rules for the use of the words 'cause' and 'effect'. In the same way, Einstein's clarification of the concept of simultaneity consisted in the precise statement of the use of this term.

It should now be clear what logical analysis achieves. It is not a procedure which 'yields' certain results, but at best a procedure which leads me to certain results – and that with my consent.

x

I should like to sum up the results of my considerations in the following propositions:

(1) Logical analysis is a way of setting out the grammar of a linguistic expression as precisely and completely as possible.

(2) The criterion for the validity of a rule in our sense consists in our *adoption* of the rule. All reasoning in which we may engage in the course of such an investigation has the sole purpose of giving us a clear view of the consequences of the laying-down of certain rules, thus facilitating our choice. But grammatical rules can never be derived from something deeper, nor can they be proved or refuted – except by other rules. The end-point of such an investigation is always the laying-down of a rule.

(3) When this kind of investigation is applied to the problem of elementary propositions, it becomes clear that the view that each proposition must be capable of being resolved into truth-functions of elementary propositions finds no support in the grammar of our actual language. We could indeed devise artificial languages constructed according to such a principle, but their structure would deviate considerably from that of the natural languages. In any language that is actually used, different strata stand out quite distinctly against one another, and the propositions of one stratum are irreducible to those of another.

(4) From this follows in particular the rejection of the positivists' attempts to 'resolve' statements about bodies into statements about sensations. The question whether such a resolution may not succeed after all by means of new logical methods points in the wrong direction, for it arises only from an unclear picture of what logical analysis achieves. In place of this question, we should ask the different question about the grammatical connection between propositions of different strata.

(5) Within each stratum we can, if we like, speak of elementary propositions, but in that case the concept of an elementary proposition will not have the absolute sense with Wittgenstein ascribed to it at one time.

CHAPTER IX

FICTION*

'Centaurs are four-legged.' If it is asked whether this is true, what ought one to reply? There is a tendency to say that such a statement is *untrue*: it refers to fabulous beings, and fables are notoriously untrue.. The contrast of fact and fiction is so firmly embedded in our whole mode of thinking that it has become hardened into a current figure of speech.

However, there is something queer about this view. If I say, in the *ordinary* sense, such-and-such is untrue, this implies that its *opposite* is true – for this is the way in which the terms 'true', 'untrue', 'false' are related. Now if I were to say, in *this* sense, that it is untrue that centaurs have four legs, the opposite of this would have to be the case – i.e. that they are not four-, but (say) two-legged. My statements, if untrue, must stand in *contrast* to others which are true. Yet this is plainly not what a holder of this view has in mind. What he wants to say is rather that statements about fabulous beings, each and every one of them, is untrue, that the whole category has to be condemned to the level of untruth.

But that only creates a new puzzle. Of two contradictory statements one is true, the other false. How, then, can both be untrue? Does the law of excluded middle break down here? A logician, nursed and nurtured in the school of Russell, will reply: That's easy enough. There is no point in insisting that centaurs are four-legged or not four-legged, any more than there is a point in insisting that the present King of France is bald or not bald – which also, at first sight, seems to run counter to logic. The mystery disappears if it is noticed that the last statement has not the deceptive simplicity its grammar suggests. In fact, there are three statements involved in it:

(1)   at least one person is at present King of France;
(2)   at most one person is at present King of France;
(3)   whoever is at present King of France, is bald.

(1) and (2) together amount to saying that *exactly* one person is at pres-

---

* Composed in English. Previously unpublished. Completed 21 November 1950.

ent King of France, and as this is false the whole conjunction is false, and remains so if 'bald' is changed into 'not bald'. That does away with the puzzle. And the same, he will add, holds of *any* statement that refers to non-existent beings – of the whole realm of fiction and mythology. If rightly analysed, all such statements will turn out to be untrue. Quite easy, you see.

This reply, it seems, gives a sensible account of the logical status of fiction. And it draws strong support from the popular view which regards fact and fiction as two opposite poles. To say of something that it is fiction is often, in colloquial idiom, to say that it is a lie – e.g. we sometimes condemn a man for untruth by branding his remarks as 'sheer fiction'. Thus the two ideas have become so intimately blended as to be almost one. But, it might be asked, is this not a light way to speak of literary matters – as if a writer of fiction was just a professional liar? And yet, has not even Oscar Wilde written an essay on the Decay of Lying, summing it up in the phrase "Art is simply a beautiful lie"? (To André Gide he once said, "I want to teach you to lie, so that your lips may become beautiful and twisted like those of an antique mask.")

This view, namely, that fiction is a tissue of lies, seems to be peculiar to this country. (I don't know about the States.) On the continent, as far as I am aware, it is far less common. What is behind the Englishman's attitude? Is it due to puritanism – with its notion that art is the arch-seducer – still lingering on in the air even after it had lost its hold on the general culture? This is not for me to decide. But I think that there is more to it than this. The English word 'true', derived from an O.E. root, which meant 'good faith', is related to *troth* ('to plight one's troth'), and *betroth, betrothal,* just as it is to German *treu.* In the oldest available records it is used of *persons* in the sense of steadfast, firm in allegiance, faithful, loyal, constant, and of *things* in the sense of reliable, sure, secure. About 400 years later, in the beginning of the 13th century, it came to be used of statements and beliefs. Yet the far older sense is still so strong – such is the inertia of language – that it deeply colours the meaning of the word to this day. Thus to ask a person whether what he is saying is true is a mild reflection on his character, insinuating that he may be lying. The deeper layer of meaning solemnly emerges in the language of the Law Courts with the sonorous phrase 'truth, whole truth, nothing but the truth'. Similar remarks apply to 'false'. The history of this word is closely connected with, and reflects, that of the Latin *falsus.* It is a word whose significance, in English, has been shaped and determined by various factors, particularly the jurists' idiom, as was *falsus* in its earlier adventures in Latin. *Falsus*, from *fallere*, originally meant deceived, mistaken, and only later took on the sense of erroneous. In modern English the sense 'mendacious' is so prominent that the word must often be avoided as discourteous whereas in German it is quite unobjectionable.

That these terms carry such a strong moral flavour in English helps us to understand – quite apart from puritanism – why the two words 'truth' and 'fiction' so forcefully exclude each other; so much so that 'fiction', like the nouns derived from 'false' ('falseness', 'falsehood', 'falsity'), approaches the sense of *intentional* untruthfulness, i.e. that of lying.

In view of these linguistic facts it is small wonder that many people, including philosophers, embrace the view that fiction is just a fabric of lies, and are content to leave it at that. Yet there is something odd about this story. To begin with, it does not do justice to the way in which we often treat fiction. Suppose a schoolboy says, 'A centaur walks on two legs, doesn't he?', then we should reply, 'No, on four'. The very fact that we *correct* him implies that there are *standards* for fictional statements, standards which we all recognize. So there seems to be a sense in which it is correct to say that a centaur walks on four legs, and incorrect to say that he walks on two.

Much the same applies to the language of the literary critics. Thus one would say, 'Yes, Polonius is the personified memory of wisdom no longer actually possessed, but a buffoon he is not'. Or, speaking of Hamlet, one would strongly object to saying that he is a coward, but would agree that he is irresolute and constantly seeks for some escape from action. Such cases, one might say, and not without a show of reason, illustrate that there is a concept 'fictional truth', paradoxical though it may sound.

There is hardly any need to pursue this line further. For there is another and much more radical objection. It is plain that, in looking on fiction as a pack of lies, one completely misconstrues it. For fiction makes *no claim* to be true; on the contrary, if a reader of a novel asks, 'Is this true?', or if a child, listening to a fairy-tale, presses the question, 'But is it really true?' we take this as a sign that he has failed to understand that it is a novel or a fairy-tale. A child has first to learn to understand a *true* story before he learns to understand an imaginary one. He discovers much later, after attaining a certain sophistication, what a thrill he can get from giving rein to his imagination, either in thinking up, or merely in following, a fanciful tale. To appreciate a fantasy requires a new technique – namely that of *disconnect-*

ing words from reality. The child has first to learn to connect names to actual objects, and he has now to learn to disconnect them and, instead, to connect them to imaginary beings. To understand, and use, names in an 'as if'-sense, completely detached from reality, is a much more sophisticated achievement, and it is precisely this which is required before one can understand fiction. The confusion should be obvious. It is not that what a novel depicts is false; it describes something *unreal* – which is very different from describing something real but describing it falsely. A statement of a witness, a balance-sheet, an excuse may be untrue, never a novel! Fiction, it is concluded, is *beyond* true and false.

So far, then, three views have emerged: – (1) all fiction is lies – e.g. to say that centaurs are four-legged is false because there are no centaurs. (2) Some fictions *have* a sort of truth – e.g. in a way it is true to say that centaurs have four legs and false to say that they have two. (3) Fictional statements are neither true nor false – for myth and poetry belong to another kingdom, the kingdom of fancy.

Each of these views has difficulties of its own. I shall take them seriatim, and try to show in each case that, while containing a measure of truth, it is not wholly acceptable.

According to the first view, it is because fictitious characters *have no existence* that statements about them cannot be true. Such a view lays itself open to a number of objections. First, and this is a point already mentioned, it ill accords with the way we often react in matters of fiction. Let us assume that, while we are talking of the fabulous figure of Polyphemus, a boy asks, 'When one of the giant's eyes was put out, couldn't he see with the other any more?' Then we would put him right. But we wouldn't tell him that Polyphemus could neither see with one eye nor with two – which would sound decidedly odd. If we did, we might, of course, assuage the boy's uneasiness by hurrying to add that the question of truth did not arise. But it seems *strained* to say such a thing, and it would be more in keeping with the ordinary way of talking to say – Polyphemus was *one*-eyed, not *two*-eyed; he was a *cyclops,* not a man; he lived on an *island,* not on the *main*land; he was *Poseidon's* son, not the son of *Zeus;* and the like. Such a case suggests very strongly that we *do* employ the usual mode of speaking to correct a child.

If we really think that all of fiction is but a fabric of lies, what, we ask, is the point of *correcting* the child, of replacing one fictional statement with another? If the people who condemn legends as lies were right, all discrimination would merge into a thick fog.

To prevent misunderstandings, it should be noticed that we are not concerned here with *quotations,* and their correctness or incorrectness. We are not concerned with the exact *wording* at all, but rather with what the words *express*; and we are apt to put a person right when he *departs* from the story. So there *is* a right and wrong in fiction.

Next, how are we to regard the *speeches* of imaginary characters? are they, too, to be carried away by the avalanche of untrueness unloosed all about them? And what of the theatre? Does the unreality of the plot render false the speeches of the *dramatis personae*? What then of the poet's true insight into the world, put into the mouth of one or other of his characters? Would not that be turning everything topsy-turvy?

But leaving aside the popular view, and turning to the way in which modern logicians deal with the matter, it strikes us how closely their whole outlook is modelled on Russell's paradigm – the present King of France. The first question we should ask ourselves is this: is it really the right move to assimilate fiction to Russell's example? We have seen how the logician was able to get out of the predicament – that of two contradictory statements none is true – by arguing that they are not *actually* contradictories. But will this line always be open to us? When Proust describes the aged and fallen figure of M. de Charlus – his hair shining with such a brilliant and metallic lustre that the locks of his hair and beard spouted like so many geysirs of pure silver – what would you make of it if he went on to speak in the same breath, as it were, of his bald head, round, and smooth, and bright like a billiard ball? Why, it's a contradiction, and as Mephistopheles says,

'A contradiction... is always for the wise, no less than fools, a mystery'.

Would you feel any better by the logician's assurance that both statements were equally false and did not contradict each other? With this the account begins to sound far less convincing.

So it does look that something is very wrong with it. Yet, it would seem, there is some force in the logician's argument. So we have still to discover the source from which spring all these confusions. I think that they spring from a fundamentally wrong conception of *proper names*, and the way in which they are used in language. According to Russell a proper name is an "abbreviated description". I shall argue that a name does not work like a description, and that this idea totally misrepresents what is characteristic of a name.

Suppose it is asked who Napoleon was, what ought one to say? There is an indefinite number of replies – the man who commanded the French army on the Nile; he who liquidated the French revolution and set up a dictatorship; the first emperor of France; the man who won the battle at Jena, or lost the battle of Waterloo; the man who divorced Josephine Beauharnais and married the arch-duchess Maria Louisa; the man who died, an exile, at St. Helena, etc. According to Russell the name stands for *a* description; but for *which* – as there are many? If different descriptions were offered to you and you were pressed to choose, you would be *irresolute*; and quite naturally so, for there is nothing which sticks out as *the* defining property of Napoleon. What is true is that, since we cannot *point* at Napoleon, we have to content ourselves with saying something which is characteristic of him. But it is *not true* that 'Napoleon', the name, stands for a *definite* description, nor that the name is being *used* in the way a description is. This can be brought out as follows. If you are asked, 'Who was in command of the French army on the Nile?', the right answer is: 'Napoleon', and not: 'The man who commanded the French army on the Nile', which leaves you no wiser, than you were before. But if a name *were* a description, it would be perfectly proper to answer in the second way. If a candidate in an examination, on being asked, 'Who was in command – ?' were to reply, 'the man who commanded –', you *would* be surprised. The point is that the question requires for an answer a *name*, not a description. 'Who lost the battle of Waterloo?' – 'The man who lost the battle of Waterloo'. Very informative. 'Who divorced Josephine Beauharnais?' – 'The man who lost the battle of Waterloo'. Glaringly inappropriate. You may easily think up hosts of replies which compete in being irrelevant, misleading, not to the point, quite out of place. The lesson to be learnt from this is that a proper name cannot be replaced by a description after a mechanical rule – by a *fixed* description. It all depends on the context, and varies with it. In this respect, such a translation, even supposing that it was possible, is very unlike a *definition:* the latter provides a *fixed, unvarying* scheme for replacing *definiendum* by *definiens*. This shows, I think, that a name has a *specific* use, utterly different from that of a description. A name is not a shorthand for a description, nor can it be unpacked into one, to use Professor Ryle's phrase. A name is a name, and a description is a description, and the one does not do the job of the other. The whole idea of 'reducing' the one to the other is mistaken, and springs from an ill-conceived attempt to transfer the technique of the scientist to a very different domain, language.

Nor would it be any use to say that the right sort of description has to comprise the *whole biography* of the man – for then *any* two sentences about Napoleon would become identical in meaning; which reduces the whole idea to absurdity. Thus a name can neither be equated to some particular *bit* of a biography, nor to the *whole* of it. Another point reinforces the argument. If we had to find out whether someone understands the name 'Napoleon', we would look and see whether he is able to use the name *correctly*, i.e. (1) as a proper name, and (2) in connection with *some* historical events which are relevant to the matter. Language leaves us considerable *latitude* in this respect (and this is of tremendous importance).

Let us, in this light, once more look at the account given for the status of fiction. The idea was to turn a name into a descriptive phrase. Suppose the statement is, 'Polyphemus was one-eyed'. Following the recipe, we shall have to paraphase the name by some such expression as 'the cyclops who was visited by Odysseus'. Let fx stand for the latter sentence-frame, and gx for the former. Then what is asserted is[1]

$$(\exists x) \cdot fx. \sim (\exists x, y) \cdot fx \cdot fy: (x): fx \supset gx$$

In words:

(1)   at least one being was a cyclops and was visited by Odysseus;
(2)   at most one being was a cyclops and was visited by Odysseus;
(3)   whoever was a cyclops and was visited by Odysseus was one-eyed.

As the first existential statement is untrue, the whole conjunction is untrue. And the same applies to any statement that occurs in mythology, in an epic, a novel, etc. *So fiction is untrue.* Q.E.D.

Now the whole argument hinges on three points: –

(a) that it is possible to replace a name by a *definite* description;

(b) that such a substitution leaves the meaning of the sentence *unchanged;* (c) that a name as used in fiction carries existential import. None of these assumptions is warranted. That (a) is not, has been shown in the foregoing; that (b) is not, can be seen as follows. Suppose we describe

Polyphemus as 'the one-eyed giant who was the son of Poseidon'.
Proceeding in the same way, we are led to analyse the statement into

(1)   at least one being was a one-eyed giant and was the son of Posei-
       don.

(2)   at most one being was a one-eyed giant and was the son of Pos-
       eidon;

(3)   whoever was a one-eyed giant and was the son of Poseidon was
       one-eyed.

As the last is a tautology, which is absorbed by the first two assertions, the
whole boils down to saying

> 'There was exactly one being that was a one-eyed giant and the
> son of Poseidon.'

By now it should have become plain that this has *not at all* the same meaning as saying, 'Polyphemus was one-eyed'. A reader who is but slightly acquainted with Greek mythology but has never happened to read the Odyssey may well understand the first sentence (which refers to Poseidon) *without* understanding the second, or if he understands the second may yet fail
to notice that both of them refer to the *same* subject – e.g. because he is un-
aware of the relation in which Polyphemus stands to Poseidon. In order to
see that both sentences speak of the same being, he has to step outside language and draw upon some extraneous source of information.

As to (c) – the sentence 'Polyphemus was one-eyed', as normally understood, contains no existential statement whatever. The sentence 'Polyphemus was one-eyed' says that *Polyphemus was one-eyed*, no more and no
less, but not that there exists, or existed, a being such that etc. It is clearly
quite wrong to suppose that the use of a proper name in fiction entails, or
implies, or suggests the actual existence of the person so named. What a
name like 'Mrs. Malaprop' evokes in us is perhaps the idea of a lady who
would speak of 'a nice derangement of epitaphs' or something of the sort,
but certainly not the idea of her *existence* – this is obviously not part of
what is meant by the name. It serves to refer to a certain character in Sheri-
dan's play, or to direct the reader's attention to it, and this is *all* it is meant
to do.

Even supposing, unjustifiably, that Polyphemus *can* be defined as the
cyclops son of Poseidon, new difficulties arise from the logician's appara-

tus – his use of functions and variables. For how is the statement, which supposedly is involved in any narrative about Polyphemus, namely, that he does not exist, to go into symbols? Is this to be expressed by writing, e.g.,

$$(\exists x, y): x \; \varepsilon \; \text{cyclops.} \; x \; \text{son of} \; y. \; y = \text{Poseidon}?$$

What, it must be asked, is the range of significance, or the *logical type*, for the variable 'y'? Gods? There are none. Organisms? But to say that a god is an organism is grotesque. 'Beings' in general? Needless to say that this is a rubber-bag category, expandible and compressible as one likes, and not precisely the sort of tool a precisian should work with; and besides, it may easily let in at the back-door some of the paradoxes which the theory of types was devised to keep out. If a logican is to be possessed by that "robust sense of reality" of which Russell speaks, we have to cast about for some means which would allow us to express, in correct symbols, the unreality of Poseidon; yet it would seem that none of the available moulds is appropriate to do the trick. So what?

But is is not only the use of *variables* that gives trouble – the whole idea of a *description* is a slippery one. Suppose I were to tell you that Chiron, the wisest of the centaurs, was turned into the constellation Sagittarius – would you understand me? In a way, you would, for you could not complain of my talking Chinese. But would you *really* understand me – in the sense, I mean, in which, in ordinary discourse, you would understand me if I told you of some episode in a man's life? You may now be inclined to say – 'Well, if I hear such things I seem to be looking through a haze or fog: – figures come and go, all larger than life, and nothing is real; the moment I try to get down to details, the whole thing melts into a cloud, and I am left with a long face'. What kind of 'description' is this then? It is all very well to say with Russell that "the sense of reality is vital in logic and whoever juggles with it, is doing a disservice to thought." But the trouble is that the mythical (which should be kept outside) already *pervades* the very description by which Chiron, or other fabulous monsters, are to be defined. So that we are like Poincaré's shepherd who, in order to protect his flock from the wolf, built a fence around them without noticing that he had fenced in the wolf as well. Are such descriptions to be admitted, I ask? If so, what will then become of'that robust sense of reality'? If not, how can the non-existence of those monsters, apparently so dear to the logician's heart, be asserted? Both roads, it would seem, are equally barred – surely a sign that all is not well with the 'theory of descriptions'?

The issue seems to have escaped most theorists. Russell, it is true, has caught a glimpse of it, but wriggles out with a few words – namely that saying 'I met a unicorn' is a *perfectly significant* assertion. He tries his best not to see the problem – and how well he succeeds! For what, in such a context, is 'significant' and what is not? Would you say that Ovid's metamorphoses are significant? Or to take an example still more drastic – what about a drama like Strindberg's 'Dream Play' where the strange atmosphere of a dream is re-captured? Here anything may happen, nothing is impossible. In the play the characters are split, doubled, and multiplied, they merge into one another like drops of water, and re-emerge. When the curtain rises, the Castle is seen slowly growing out of the soil like a giant flower. The person who is Agnes, Indra's, and at the same time the glazier's, daughter moves through the scenes with her husband, the lawyer, who changes into the officer, and then the poet, who are all the same person, and again are not. In the end, the Castle rises up in flames, and in their light a wall of human faces is glimpsed, questioning, sorrowing, despairing. Now if it is asked whether this makes sense, what ought one to reply? It looks as if we understand many things which, in a prose-way, we would not understand. Or to put it differently, myth and poetry allow us much more *latitude* than ordinary discourse: what is enough for mythology is not enough for home-made descriptions. We can afford to be far less particular in the former case and to dispense with all the credentials we insist on in civic life. Or again, what is meaningful in fiction may lose its meaningfulness when taken in deadly earnest. Our standard of what a description is varies *enormously* with the kind of statement before us, or rather with the context which is made part of. And so we see that the concept of description fades imperceptibly into a vague twilight zone where we can no longer confidently say, 'this is a description', or 'this is not a description'. And we notice at the same time another curious point – that the whole array of allied notions – 'statement', 'significant', 'meaningful', 'intelligible', etc. – undergo a parallel change in sense. Incidentally, that is one of the reasons why the tribe of criterion-hunters – with their battle-cries 'verification!', 'wrong!', 'falsification!', 'none of that!', 'logical syntax!', 'better still – semantics!' – are on the wrong track. The point is that all these terms have an *ambiguity,* an inevitable one, which foils any attempt to pin them down. The logican, guileless as he is, believes that there is one mould only into which each of them is to be squeezed; well, and this results in bruises. Meaningful, meaningless – if

I always knew exactly what I was talking about, what pleasure, I ask, would there be left? Our actual situation is rather more like that of the man of whom an 18th century author said

> And now to sense and now to Nonsense leaning
> He stumbles on and splutters out a meaning.

And now, I suppose, I may go on with the job.

A new complication arises, in that fact and fiction not seldom permeate one another. Odysseus is not *only* a mythical being – behind the huge shadow a historical figure is faintly glimpsed, the figure of an adventurer, sailor, and king. And, indeed, there is a kernel of solid truth in the whole legend of the Trojan war. How is one to say whether there really *was* an Odysseus? That may depend on what sort of adventures one attributes to him. Suppose, for instance, you were asked, 'Would you say that Odysseus had existed if there actually *was* a man living at the time of the Trojan war who was called "Odysseus" by his contemporaries, but of whom nothing else Homer narrates was true?', you would presumably reply in the negative. On the other hand, if it could be established that, in fact, there lived a man at that time of whom everything recorded in the Odyssey was true except that he was not called 'Odysseus', you would be inclined to say that he was a historical figure. Between these two extremes a vast number of alternatives can be thought of, in the sense that there was a man who did *some* of the deeds ascribed to the Homeric Odysseus, though not *all* of them, and who was, or was not, known by this name. It is easy to see that on one interpretation Odysseus may be a mythical figure, on another an historical, with the result that a statement like 'Odysseus returned home to Ithaca' may turn out to be true, or untrue, according to the description one chooses. But *is* one to choose? And how? The trouble is that the legend itself is a sort of centaur, half fact, half fiction, but so intimately joined together that you cannot, with a scalpel, dissect the creature and neatly sever what is historical from what is decent myth.

You may feel here a need for the use of some such terms as 'half-myth', or 'half-fiction'; but the snag is that you would hardly say of a statement of half-fiction that it is half-true. And in the modern Odyssey, too, Stephen Dedalus and Leopold Bloom, fictitious characters, are placed in a *real setting*, that of Dublin, as it stood on Thursday, June 16th, 1904. The narrative of the episodes, thoughts and dialogues, all of them imaginary, is everywhere interwoven with references to actual things – the town and its local landmarks, the people of Dublin (who are called by their real names), political incidents of Ireland at that time, and so on. The whole story is saturated with allusions of this sort. As a consequence, both ingredients, the fictitious and the factual, may interpenetrate in one and the same sentence – as when Bloom is depicted walking down a Dublin street, and talking to one of its actual denizens. Thus the result is the same as before: the question 'fact or fiction?' cannot always be resolved, anyhow not without some arbitrariness.

None of these difficulties is raised wantonly – they force themselves upon anyone who tries to *understand* what the status of fiction is. It is for all these reasons, I submit, that the philosopher's equation 'fiction = falsehood' must be rejected – a result anticipated by Mr. Hare when he wrote, in 1827, that "the facts of Poetry, being avowedly fictitious, are not false."

If fiction is not false, ought we, then, to say that it is true? If we take ordinary language as our guide, there is little doubt that we sometimes do express ourselves in this way; (cases illustrating this have been cited before). To return to our old example – there certainly seems to be a sense in which it is 'right' to say that a centaur walks on four legs, and 'wrong' to say that he walks on two. Besides, how have we come to understand such a word? Perhaps we have been told that centaurs are beings half-man and half-horse, perhaps we have been shown a picture of such a being – Böcklin paints the most convincing centaurs and mermaids. Thus we may recall a verbal explanation, or a pictorial one. Should one not rather say, then, clenching the point, that such a statement is *indisputably* true – namely *analytic,* being grounded in the idea, or the image, we make ourselves of a centaur?

But this surely is pushing the matter almost to the point of absurdity. Granted even that there is an atom of truth in this special case, the example is *too* special. No one in his senses will try anything of the sort in the case of any old character in a novel. This, however, leads us to another point. Generally, one does not apply the notions of true and false to the realm of fiction. One does so only in *very special* circumstances – when a work of fiction, or poetry, has gained general currency, when it has, in fact, become part of the national heritage – such as the Bible, or Shakespeare, or Milton. It is only when talking of characters that have attained a certain recognised status that it is proper to use the idiom of true and false. To apply it to the creations of some obscure third-rate novelist would be ludicrous. Here an entirely new factor makes itself felt – the accepted *standard* of a literary work. The more it has become part of the living heritage of a people, the more is one prepared to employ this idiom.

If we do employ it, the word 'true', when used e.g. to refer to something in the Bible, can mean either of two things – true in the *ordinary* sense (the Flood did really happen, or true *within* the Bible. In the latter sense it would be 'true' to say that the Flood was sent by the Lord as a punishment for the sins of man, just as, in a different frame of reference, it would be 'true' to say that Achilles avenged the death of his Patroclus.

However, there is a more fatal objection to all this, namely, that fiction is sometimes *intended* to be lies. The classical example of this is Münchhausen. If we call the Baron's tales lies, it is not because these adventures never really happened to him, but far more because of their utter incredibleness, and complete disregard for the bounds of the possible – his frozen and reanimated trumpet note. What lends these stories their charm is the completely matter-of-fact tone of the Baron's narration, who speaks as though these unbelievable adventures were the merest everyday incidents of his life. This extreme mendaciousness of Münchhausen's, together with the considerable currency which his tales have achieved, can lead to a curious situation. Thus if someone narrates a Münchhausen adventure, and narrates it wrongly, he might be challenged 'No, it was not so', and then corrected – just as with the centaurs. Thus one could say, 'Yes, *that's* what is true' (namely within the frame of the story), and yet agree what the whole is nothing but a pack of whopping lies. Here a sense of 'true' neatly cuts right through a sense of 'false' – without necessarily leading to a contradiction. Incidentally, it is debatable whether the Baron really is the model of an archliar, considering that his tales are so glaringly untrue that they can deceive no one; from which it would seem to follow that a lie, if only big enough, ceases to be a lie.

Thus the tenet that fiction holds a truth of its own is untenable too. 'But if fiction is neither true nor false, what is it?' Well, it is just what you said – neither true nor false. What makes this view so attractive is not only that it offers us an escape from two evils, but also that it seems in itself more appropriate to the nature of fiction. The world of fancy, it is often said, is a world of its own where truth and falsehood may stake no claim. Such a view may, perhaps, be underpinned as follows.

An ordinary statement of fact always contains two parts – one that is *pointing,* and one that is *descriptive.* Thus if I say, 'I am sitting on this chair', what is pointing is the word 'I' and the word 'this', and what is descriptive is 'sitting on a chair', and when I say, 'At 20A Banbury Road there is a red-brick house', the address serves to identify a particular portion of reality. Any language contains, must contain, certain words for fastening a descriptive phrase on to a piece of reality – e.g. the pronouns 'I', 'you', and 'he', (the last suitably explained, e.g. by pointing), the adverbs 'here' and 'there' (with an accompanying gesture), 'now', etc. Another group of indicator words are proper names. Now what happens in fiction is this – the pointing parts, amongst them the proper names of the characters, are only dummies: they *pretend* to point, but they do not. *Therefore* the story is not false. If the names *did* point, and if the episodes narrated in it turned out not to have happened, *then* it would be untrue. But to construe it in this way is to *mis*construe it. And even if a name, e.g. Henry VIII, does refer to a real person, it is not *meant* to do so in Shakespeare's play. By the way, fiction employs certain devices for making it clear that it is *fiction* – unrelated to the real world, that is. Thus the phrase 'Once upon a time' may be likened to a *clef* in music: it sets the tone, it explains right from the beginning that what follows is a fairy-tale. It is this unconnectedness with reality which makes it pointless to talk of fiction in terms of true and false.

This sounds plausible enough. Yet you may feel slightly uneasy about swallowing this account. But why? What you are not happy about is perhaps this. Our concern, you may say, is with *statements*, and statements, as opposed to names, are by nature 'bipolar' (to use an expression of the earlier Wittgenstein); that is, statements lie in a sort of field suspended between the two poles labelled 'true' and 'false', or '0' and '1', or 'α' and 'β'. (In parenthesis, what is the difference between 'fire' when I mean it as a *word*, and 'fire' when I mean it as a *statement*? Simply this that, in the latter case, it has a specific use – a use, that is, which makes it meaningful to apply to it the terms 'true' or 'false'. It is made part of a system of operations, of a two-valued calculus. It is this which explains why a sentence is not a sort of name given to a fact, a label tacked on to it: if it were, the question of truth would not arise.) To say now of statements which occur in a novel that they are neither true nor false strikes at their very *essence,* the definite usage which makes a statement a statement and distinguishes it from a mere name. To accept such a view would commit us to say that fictional descriptions are not descriptions. How, then, can they tell us anything, if only something imagined? Here is the rock where this account of the matter, too, is shipwrecked.

I have given now the reasons why, in my opinion, none of the three views I have so far enumerated holds. The 'theory of descriptions' held in high esteem by logicians, has proved a mine of mistakes; nor can any of the other accounts be defended – we cannot go on saying that fictional statements *have* a sort of truth and yet cannot be true, nor can we embrace the third horn of the trilemma, the view, namely, that fiction has nothing at all to do with the notions of truth and falsehood. It is now time, I think, to ask ourselves – what is at bottom of all this?

To turn to the positive account of the status of fiction – I think the way out is this. Descriptions as they occur in a novel are, indeed, bipolar; for otherwise there would be no logic in a narrative and everything would become blotted out (despite an heroic observation of Mr. Herbert Read's that "literature, poetry in particular, can dispense with logic"). But if we ascribe to them truth or falsity in the *full* ordinary sense of the words, we are already setting one foot on the wrong track; since this commits us to recognize them as ordinary statements, i.e. to do the very thing which we have to avoid, and which in fact, we had to give up as hopelessly mistaken. Yet to deny them *any* relation whatever to those notions will not do either – as this in effect would mean to rob them of the nature of descriptions and perhaps, reduce them to the level of names. Thus we have, it would seem, to mediate between two conflicting claims: on the one hand, we have to concede to them some measure of truth and falsity, on the other hand we cannot allow them parity with ordinary statements. What we have to discover is how this can be done.

I want now at this point to introduce a distinction betwee the *formal* and the *non*-formal, or *material* features of the notion of truth. It may help to explain the issue if I first draw attention to a somewhat similar difference with which we are all familiar. Take the concept of equality. It is a relation that has a certain 'formal structure', and this structure can completely be stated by saying: the relation 'equal' is reflexive, symmetrical, and transitive; in symbols: –

(1) $a = a$;

(2) $a = b. \supset . b = a$;

(3) $a = b. b = c. \supset . a = c$

Any relation which possesses these three formal properties can be said to be an 'equality relation' – for instance, equality between numbers, or congruence between figures in geometry, etc. That is, we would not recognize a relation as an equality relation unless it satisfied these three conditions. But is this enough to enable us to employ the term in practice? Obviously not: if you know what it means to say that two bodies are equal *in length*, you do not yet know what it means that they are equal *in hardness*, or equal *in temperature*, *in weight*, etc. Nor are you, after you have learned to compare intervals *in space*, any wiser as regards comparing intervals *in time;* and it is precisely on account on the *unlikeness* of comparing that St. Augustine could have been mistaken, when he believed that measuring of time must be *exactly like* measuring of space and, as a consequence, sought desperately for something like a temporal tape-measure, a 'yard-event' which can be carried back and forth through time like a yardstick, and applied to events in the past or the future. In short, our three formal conditions in no wise *exhaust* the full meaning of the word 'equal'. What we have still to know in order to be able to use it intelligently, i.e. in different sorts of situations, with regard to length, duration, temperature, and so on, is a *method of comparison*. In other words, what has to be added to the rules is something which is *not* formal, but material, in the sense that it supplies us with a *criterion*, or a *test*, for deciding the issue. This test varies with the sort of subject to which it is to be applied. Compare 'the same distance' (in space), 'the same interval' (in time), 'the same hardness', 'the same temperature', 'the same force', 'the same brightness', etc. Notice in which respects they are alike and in which respects they are unlike. What they have in common is the mere skeleton of the concept (as stated in our rules), what is different is the criterion of equality, – scratching in the case of hardness. Hence the term 'the same', or 'equal', is each time used in a *different sense*, owing to the specific technique to be applied. And it is this circumstance, namely, the need of adding differing criteria suited to different classes of properties to the mere shell of the concept, which generates a variation in meaning – a *systematic ambiguity*.

It seems to me that one can, in a very similar manner, speak of certain *formal* features of truth, as opposed to others which are *material*. What is formal in the notion of truth are those things which matter in logic. What is required for the purposes of logic is, *au fond*, only an *abstract* division of statements into two classes 'true' and 'false' from which any relation to the *content* of these notions has evaporated, and further the fact that the transition from the one class to the other is governed by certain rules and operations. Thus it is a *formal* feature of truth that the negation of a true statement leads to a false one, that what follows from a true statement is true, etc. A *non*-formal characteristic of truth is expressed by saying that the kind of truth a statement has varies with the kind of statement it is.

We are now in a position to see that there is a parallelism between the notion of equality and that of truth, as far as the relation of 'formal' and 'material' is concerned. Indeed, what corresponds to the method of comparison is the *procedure of checking* a statement, i.e. the method of *establishing* its truth. In this respect different kinds of statement behave very differently – for instance, a plain statement of fact, or the ascribing of a specific motive to a person, a physical law, the theory of evolution, or what not – (not to mention such things as a proverb, or an aphorism, which raise peculiar questions of their own). Now the way in which truth is established in each of these cases, and whether it can be established at all, conclusively or not, has a bearing on the notion of truth itself, and changes and modifies it in ways which deeply colour its signification. Accordingly, the term 'truth' acquires a multiplicity of meaning, just as the term 'equality' does. And the reason for this is the same in both cases, namely, that only *part* of the full and rich content of these notions can be formalized, whereas the other one, the material, serves to *adjust* the abstract notion to the varying needs of reality.

It is in this way – detached from reality, unconnected with any method of checking, in short, *free* – that bits of narrative must be regarded. And it is in keeping with this that only a pale residue of what we normally include in truth and falsity is preserved – yet enough to guarantee the smooth working of logic; and it is also *this* which secures them the character of *descriptions*.

From this we get a glimpse of the logic of poetry which reveals it as something curiously like the logic of mathematics. For the mathematician, even more than the poet, must cut himself free from reality. "The essence of mathematics lies in its freedom" as Cantor said. In both mathematics and poetry language is cut off from that connection with reality which the normal use of indicator words establishes. In both cases the concept of truth is, as it were, stripped of its content, though in differing ways and to differing degrees. A historical novel is closer to the world than mythology, and ordinary geometry closer than abstract algebra. Only the mathematican is more radical in that he divests his symbols even of that descriptive content which is essential to fiction. But what both domains have in common is that contact with reality is not only not sought, but purposely avoided; yet in both cases the formal bipolarity, so essential to statements, is retained. And it is this fact which explains that even in the highest flights of imagination we are still bound to respect the laws of logic – that even mythology dares not go so far as to say that Polyphemus was, and yet was not, Poseidon's son.

NOTE

[1] [Waismann here employs a notation like that suggested in Wittgenstein's *Tractatus* 5.53 ff, in which there must be distinct substitution for distinct variables.]

CHAPTER X

A NOTE ON EXISTENCE*

Would it not be very queer if a man pointed with his forefinger at something and said 'This exists'? Of course it does; so why say it? What is he up to? I do not want to deny that a situation can be imagined in which words to this effect would be to the point. Suppose, for instance, that I am in the desert and someone says to me, 'Look, what you see yonder on the horizon is a mirage – it does not exist; but all that' (pointing around) 'does'. Exceptional cases apart, no one would say this. What one says is, e.g. 'There is another chair in the next room', 'Nebulium exists in gaseous nebulae'. Even the assertion 'There *exists* another chair in the next room' sounds somewhat stilted; but to assert of the chair on which I now sit that it exists is, to say the least, nebulous.

But what exactly would such a form of words express? If 'This exists' is symbolized by writing '($\exists$ x) x is this', should we construe this as saying 'There exists at least one thing in the world which has the property of being this'? That seems a far-fetched interpretation, reminiscent of the *haecceitas* of ill repute. So what can it mean? Consider first another point. If it is correct to say of a thing at which I now point 'This exists', it is also proper to say 'This does not exist', or 'There is no such thing that is this' – certainly very puzzling phrases. Indeed, what can I mean by denying existence to something that stares me in the face – in normal circumstances, that is, not being faced with a mirage, etc.? We somehow feel this contrary to sense, no less than saying 'This is not there', where the word 'there' indicates the place of the object I am pointing at. Not there? Where on earth should it be if not where my finger is pointing to?

It will be convenient to use arrows to stand for the act of pointing. An arrow occurring more than once in a given context is to be understood as standing for the same act of pointing. And by 'the same act' I mean the same *physical* act, although the intention may each time be different. Then the foregoing, namely, 'This is not there', may be expressed in the form 'This →

---

* Composed in English and dated July 1952 on the reverse of the first page of typescript.

is not there → ', thus bringing out more forcefully its nonsensicality. More-over, it may be observed that the latter expression is closely akin to 'This → exists' – a locution which to all appearance may be rephrased as 'There → is a thing such that it is this → '; which, in its turn, seems to amount to saying 'There → is this → '. But surely saying

(α) 'There → is this → '

is just as without rhyme or reason as saying

(β) 'This → is there → '?

Before taking up the point, let us see in what relations the two expression stand to one another. If they are independent, it must make sense to assert the one while denying the other, i.e. it must make sense to say of anything at which we can point,

'There → is this → , though this → is not there → ' and 'This → is there → , though there → is not this → ';

which, according to the suggested usage, would retranslate itself, respectively, into

'This → exists, though it is not there → ', and 'This → is there →, though it does not exist'

– turns of phrase which sound very odd, if not contradictory. If they *are* contradictory, or if we *declare* them to be so, this would mean that the expressions (α) and (β) entail each other, and if they entail each other, they mean essentially the same thing. Now if 'This → exists' has the same sense as 'This → is there.→ ', it has *no* sense – it degenerates into a sort of tautology; which, after all, is not so surprising.

It will be noticed that the argument loses none of its force if the expression 'This → exists' is paraphrased in a slightly different way, namely, (leaving out the one arrow) by writing 'There is a thing such that it is this → ' – which comes to the same as 'There is this →'. Our conclusion, then, is unaffected by the special form of paraphrasing we adopt.

But is 'This → exists' tautologous? If so, its opposite, i.e. 'This → does not exist' must be a sort of contradiction; which sounds rather implausible. By making the most of a play on words and turning 'There is this → ' first into 'This → is there', then into 'This → is there → ' and thus into something *openly* tautologous, we perhaps were too closely following the clues of speech. Anyhow the phrase has a slightly tautologous ring; whether it really is tautologous may for the present be left in suspense.

Some clues, however, tell a different tale. If it is correct to express 'This → exists' by writing

(1) $(\exists x). (x$ is this $\to).$

it must also be correct to say

(2) $\sim(\exists x). (x$ is this $\to);$

in words, 'There is nothing which is not this →', which, according to the customary rules of symbolic logic, can be rewritten as

(3) $(x).(x$ is this $\to),$

i.e. 'Only this → exists', 'Everything in the world is this →', 'This → is all there is', etc. This is patently absurd. Yes as these bits of nonsense are obtained from (1) by carrying out two perfectly lawful steps – negating the given expression twice, once in front of the prefix, the other time within the sentence frame – we can only conclude that our starting point is equally nonsensical. These symbolic operations have an effect like an amplifier – they magnify the initial bit of bosh so that no one can fail to notice its nonsensicality.

In a shorter way, the same result can be obtained without having resource to the symbolic apparatus. If 'This → exists' is significant, there is no reason why 'Only this → exists' should lack significance. Does this mean that there is nothing which is not 'this →', or that 'this →' fills the whole universe to the brim? But in that case the word 'this' and concurrently pointing at something only serves a purpose if I could also have pointed at something else, i.e. if the pointing singles out, or brings into prominence, some particular item – if it contrasts something with something else, even if only with an undifferentiated background. But if 'this' applies to everything indiscriminately, it fails to perform its function. (A sidelight on how to get rid of a similar confusion, solipsism.) But, it might be objected, surely I can imagine that there is only one thing in the world to which I refer by the word 'this' so I can give it a use. But let us see what would it be like if my pointing was no longer selective? What *am* I to imagine? Perhaps just a single star on the sky, and nothing more? Needless to say that, in this case, there would at least have to be some expanse of dark space around it which would also be a suitable 'something' to point at or call attention to. Besides, if what I was supposed to point at was *everything* there is, how could I even *point* – if neither I, nor my pointing finger existed? This vitiates the attempt to give sense to the phrase 'Only this → exists'; and as sauce for the goose is sauce for the gander, the argument applies equally to '*This* → exists'.

The point is that we neither say 'This → exists' nor 'This → does not exist'; odd cases or jesting apart, we have no use for either of these locutions, nor do we wish to give them a use. But surely, our objector might resume, there must be *some* sense in saying 'This → exists', – for the following reason. Imagine a situation, such as described by Andersen in *The Emperor's New Clothes*. His Majesty walks on the street with nothing on, and all the spectators cry, 'Look, how splendid this → is, how magnificent that →!' Yet there is nothing answering to these words and gestures. Here, then we have a case in which it seems proper to say that the things referred to in this way are unreal – that had the people spoken the truth, they would have said, 'All these → things' (relating to the Emperor's clothes) 'do not exist'. So we would use one of these very locutions. But to say that what the people point at is non-existent is an atrocious way of putting it – there is always something in the straight line of the forefinger when a man points. The truth is that the people do not really point at anything, but only pretend to; i.e. they behave as if they were pointing at something, while in fact they are not. And the difference between real pointing and sham pointing would become clear at once when they were asked to detail what it was they were pointing to at this very moment and which, to all appearance, filled them with such delight.

That goes far to resolve another curious puzzle – can one point to something imaginary? Suppose that I am a playwright, and, fired by my own imagination, 'see' the characters of my play standing around me: beaming and laughing I enter a gesticulating conversation, saying funny things, complimenting the one on his success, pointing at another one and chaffing him on his state, making repartees to imagined questions – in short I treat them as if they were present in the flesh, and while doing this, for the moment almost believing in them. How are we to account for that? The answer is clear: I do not really point, I only profess to do so – though there wouldn't be much difficulty for me in describing them to a *T*.

There is a tendency to say that, if I am pointing at something, 'really' pointing and not merely shamming, the object of my pointing must exist. Certainly; but what the objection overlooks is this. To say that I am pointing is not to point, and to say that the object of my pointing exists is not to say that '*this* → ' exists – although it is true that the two expressions, 'the object of my pointing' and 'this →', have the same reference. There is nothing more astonishing in this than in the fact that the word 'I' and the proper name 'F. Waismann', though they both refer to the same person, are not necessarily interchangeable. Thus in saying 'I am Waismann' to ontroduce myself to a stranger, I am not saying 'I am I', nor 'Waismann is Waismann', any more than saying 'Today is Friday' is just another way of saying 'Today is today', or Friday is Friday'. And just as, at least in some contexts, 'today' cannot be replaced by 'Friday', or the first personal pronoun by my name without marring the sense, so the wordt 'this', accompanied by a gesture, and the phrase 'the object of my pointing' cannot be substituted for one another. The latter cannot be used in the way the former is, i.e. the expression 'the object of my pointing' cannot be made to act as pointing; it does a bit of describing, but 'not a bit' of indicating.

What has been said seems to favour a somewhat different version. Uttered in the presence of an object, one might say that the phrase 'This → exists' is pointless, absurd – stressing its incongruous rather than its tautologous aspect. Yet it is not absurd in the sense of being unintelligible. Supposing that someone were to tell me that he is looking at me with his ears, or hearing me with his eyes, I should not understand what he was talking about. 'This → exists' is different, though we cannot help feeling that there is something fishy about it, yet we seem to have no difficulty in understanding what it says. The expression is not so absurd as the illustrations in the previous sentence nor as tautologous as the examples in the foregoing paragraph. Nevertheless it possesses an inappropriateness the cause – or ground? – of which we shall shortly have to consider.

Since a certain argument proposed by G. E. Moore is in direct conflict with the above observation, we must examine it at this juncture. He writes[1]:

. . . to point at anything which you see and say 'This exists' seems to me to be meaningless . . .; but I cannot help thinking that in the case of anything to point at which and say 'This is a tame tiger' is significant, it is also significant to point at it and say 'This exists', *in some sense or other*. My reason for thinking this is that it seems to me that you can clearly say with *truth* of any such object 'This *might* not have existed', 'It is *logically possible* that this should not have existed'; and I do not see how it is possible that 'This might not have existed' should be true, unless 'This does in fact exist' is true, and therefore also significant. The statement 'It is logically possible that this should not have existed' seems to *mean* 'The sentence "This does not exist" is significant'; and if 'This does not exist' is significant, 'This does exist' must be significant too.

This looks at first and second sight very persuasive. I shall, however, try to show that the argument is nothing but a logical muddle, and I shall do this in two steps. I shall argue first that Moore's test fails in cases which, to all appearance, are perfectly similar, and then that in any case the validity of his argument is more apparent than real. In fact it is this show of argument which contributes the element of persuasiveness. It is not based upon any logical pattern of reasoning, any 'form' that could be extracted from the linguistic garb and exhibited in logical purity. Like Descartes' move – some theorists prefer to say 'inference' – from the fact of thinking to the presence of an agent that thinks it owes its strength to numberless analogies in language. Thus we say 'True, he might not have come, yet here he is', 'He might never have met her, but fate decreed it otherwise', 'I might have died, yet I'm carrying on'. And just as 'He might not have come' does not make sense, unless 'He did in fact come' does, so we are apt to pass from 'This might not have existed' to 'This exists', even where the object referred to is just in front of our nose, and to argue from the fact that the former expression is significant that the latter is. In this, we merely follow the channel grooved out by language; and the force we feel in the argument is the weight of all such similar instances. It is a force not logical, but analogical.

But when I say that we are guided by analogy I do not suggest that this is always a conscious process. Usually it is not. Rather, we seem to to be sensitive to them in the way, perhaps, in which we are sensitive to a face, or tune which strikes us as vaguely familar. It is as if the presence in language of such analogues, however inarticulate and hazy, exerts upon us an influence almost like a magnetic field – a kind of language field – gently impelling our thought along certain lines. Thus we are almost irresistibly induced to argue as Moore does, supported as the inducement is by the whole weight of language – notwithstanding the shadowy character of the parallels. And, I venture to say, it is precisely because of their fleeting, half-formed nature that their influence is so pervasive.

But they may play us false. Indeed, if the argument were firmly anchored in logic, it would be monstrous even to think of counter-examples. As it is, it is not at all difficult to produce some. But before turning to this part of my programme, an error must be corrected which unnecessarily complicates and obscures the issue.

It will be remembered that Moore operates with the term 'true' in a somewhat peculiar way. He says: "I do not see how it is possible that 'This might
not have existed' should be true, unless 'This does in fact exist' is true, and
therefore also significant". True? Can it be verified? And if so, how? What I
can verify is that the thing in question exists, but not that it *might not* have
existed. Why, then, insist that one may say '*with truth*' of any object one
may wish to pick out that it might not have existed? What, exactly, does the
term 'truth' mean here? Since all Moore is concerned with, and all his argument is meant to establish, is that saying 'This → exists' as well as saying
'This → does not exist' is *meaningful*, the question of truth is beside the
point.

Should any doubt remain, consider the matter from a different angle.
Suppose that, sitting in my armchair with the blinds drawn, I say 'It might
be raining'.[2] Well, if it is not raining, is my statement false? Can it be a lie?
This presupposes that I *know* that it is not raining and deliberately say the
untruth. Was I, then, mistaken in point of fact? If I had said 'It *is* raining',
i.e. if I had committed myself, then I would have been mistaken. But I have
guarded myself against this reproach by using the form 'might'. In fact,
'true' and 'false' are inappropriate in this context; (which, by the way,
shows that it is a prejudice of logicians to believe that every statement *must*
be either true or false). The relevant question here is whether that was the
'proper thing' for me to say; and this depends entirely on the whole situation in which I uttered the words – whether I had any reasons for believing
what I was saying (a pattering noise on the window-panes, the weather
forecast, a twinge of rheumatism, and what not), or whether I was talking 'at
random'. Notice that a whole tribe of idioms are employed in a similar line
– 'It's perhaps raining', 'It's probably raining', 'It sounds as if it is raining',
etc. To approach any of its members with the question 'True or false?'
would be strangely inappropriate.

The case 'This might not have existed' differs from the previous ones.
Here I am not talking guardedly; on the contrary, I clearly imply that the
thing exists. To say of something which does not exist and of which I know
that it does not exist that it *might* not have existed would be not false, but
pointless; and if the thing does not exist, though my remark may now be to
the point, it is not true. Whether apposite or not, in no case am I *asserting*
something, any more than I can devise an *experimentum crucis* to put it to
the test.

But what about saying 'It is *logically possible* that this should not have existed'? Is this at least true? In a sense, it is. But then we are talking about the way we are talking, (though, mind, not about our speech habits), and the whole notion of truth is now a different one. For no longer pertaining to facts, it cannot be construed in the classical sense of 'accordance with fact or reality' – evidence that the forms 'This might not have existed' and 'It is logically possible that this should not have existed', though they *almost* come to the same, do not *exactly* come to the same. By 'translating' the one idiom into the other, something has happened to the one, with a bit of sense-shifting on the way. Indeed, saying 'It is logically possible that this should not have existed', has an *assertive force* which is lacking in the former, and this force is even more apparent in 'The sentence "This does not exist" is significant'. The change is quite conspicuous in the case of 'It *might* be raining' which is altogether different from 'It is *logically possible* that it should be raining'. But there is *no sharp line* separating the two kinds of idioms. Compare, for instance, 'It is probably raining' with 'It is probable *that it is* raining'; you notice that the latter, though *less* noncommittal and, in virtue of the 'It is that'-construction, *more* like an assertion, yet has not the *full assertive force* it would have had I said 'The probability that it is raining is such-and-such', giving an exact figure. (Plenty of room for more casuistry.) From this it is seen that the two types *shade off* into each other – with the corollary that we should be unrealistic, and fail to do justice to the actual forms of reasoning, if we drew too sharp a line between the 'material' and the 'formal modes of speech'.

But to return to Moore – the question of truth, it has been shown, is beside the point. His argument, therefore, wants pruning, and when we do so it may be re-stated as follows. Since 'This → might not have existed' clearly makes sense, 'This → does in fact exist' must also make sense; and if 'This → does in fact exist' has significance, then the same status must be accorded to 'This → does not exist'. (This argument, needless to say, is again backed by countless analogies in language. Hence its convincingness.)

Yet it is fallacious. For example, it makes perfectly good sense to say 'I might not have existed', namely, in the sense that I might never have been born. But does it follow from this that I should be talking sense if, pointing at myself, I said, 'I do not exist'? This would not exactly strike us as a paragon of clear reasoning. Or suppose I were to argue: What is here – namely the pencil lying on the table in front of me – might be somewhere else, say, in the drawer. Would it make sense if, following a somewhat similar line, I argued that 'What is here is in fact here' is *significant*, and now went on to the bitter end, concluding that it must therefore also make sense to say 'What is here is not here'? What has gone wrong? One last illustration. No one is in doubt as to what I mean when I say, somewhat wistfully perhaps, 'I might have lived in the past'; and there is indeed not the least difficulty in imagining myself as living at some former period of history. What now will become of our argument? Imitating Moore, as far as possible in his own words, I should have to say something like this. I do not see how it is possible that 'I might have lived in the past' should be significant, unless 'I am in fact living in the past', (namely now) is also significant. The statement 'It is logically possible that I might have lived in the past' seems to *mean* 'The sentence "I am now living in the past" is significant'. Yet the last sentence is plainly absurd, (unless when taken in a figurative sense which, of course, is not in the line of the argument). What has gone wrong?

What has gone wrong is this. It is quite correct to say of a thing that is before us, (say) of the pencil, 'This might not have existed'. But what do we mean by this? Do we mean that *if* I point at the pencil, it might all the same not exist? Clearly not. What we mean is that it is *logically possible* that I should not have pointed at a pencil such as I see now, and that, for all we know, an object like this should not exist, and/or should never have existed; i.e., we can easily imagine a world from which a thing like this particular pencil is absent. But once I *point at it* with my finger, I am no longer free to say of what I am indicating that *it* does not exist. If I were to say this I should be making the futile attempt to undo what I am doing. Instead of saying 'It is logically possible that such a thing might not have existed and that I should not have pointed at anything like it', which would have been perfectly correct, Moore insists that it *makes sense* to say that 'this, as singled out by my ostensive act, does not exist'.

What he has failed to see is the part played by the two phrases, 'This → might not have existed' and 'This → exists', and the way in which they are related. He wrongly supposed that, just as 'He might not have come' says by implication 'He did come', so 'This → might not have existed' says by implication 'This → does exist'; i.e. he construed these two phrases as being related in the same way as the other two are, thus following the lead of analogy. Now we may, it is true, use the phrase 'This → might not have existed', but if we use it we do not utter it as if we were proclaiming some deep and

exciting discovery; we utter it in a matter-of-fact tone, just as if we were uttering a platitude. Indeed, what more do we wish to express by this phrase
than that everything that is might be otherwise, that facts are what they are,
contingent? It is nothing out of the ordinary, no piece of Eastern wisdom we
are driving at. We merely call attention to a notorious feature of ordinary,
empirical facts and nothing more – very much in contrast to 'This exists'
which may give us an important piece of information. And even when it
does refer (say) to the pencil that lies just in front of our nose, 'This → exists', if the phrase is used at all, is used in a curious sort of way, namely, as a
way of *ending a dispute*. 'Granted, it *might* not have been there', we say,
'but there → it is – can you deny it? It is like saying 'That's that', a formula
closing discussion, or implying 'It's not good arguing any further'. It is in a
similar sense that our phrase clenches the point. Compare also 'A penny is a
penny', 'I am what I am', 'War is war' – idioms not meant as exercises to illustrate the 'law of identity'. Take the sentence 'A penny is a penny': a remark made only in certain circumstances, and, further, in a certain tone of
voice, intimating that you should not disdain a penny, that you do not find
it in the street, and the like. There is a certain *point* in saying it; it is not the
truism it sounds. 'War is war' – a formula used to drive home some such
thing as: You see, *that* is what war is like; in war such things *do* happen; after all, war is not peace; what can you do? there's a war on; etc. As with 'War
is war' and 'That's that' so with 'This exists': *it is not pointless*; on the contrary the speaker wishes to drive home some particular point when he utters
the sentence. It is, then, the context which gives point to it. Taken in isolation, the phrase is what it is – *almost* tautologous, *almost* absurd; and logically iridescent.

It will by now have become clear what I am aiming at. 'This might not
have existed' is related to 'This exists', not as 'There might have been no
war' to 'There is a war on', but rather as 'There might have been no war' to
'War is war'. That, perhaps, is a bit stretching the point, for no parallel is
exact, but stretching it for didactic purposes. Ordinary language is full of
tricks and traps, just as it is full of meat and wisdom. Moore, having failed
to notice the widely different ways in which these two phrases are used, and
the widely different circumstances under which they are appropriate, was
led to take them literally and look for a sense where none is to be found.

That this is the core of the matter can, moreover, be seen when we try to
apply Moore's argument to the sphere of mathematics, where we do not

have to reckon with contingency. Suppose I said 'There might be no prime numbers', does this offer us any guarantee that saying 'There are in fact no prime numbers' is meaningful? Or suppose I argued: There might have been a fourth dimension; therefore it must make *sense* to say that our space is in fact four-dimensional? The case may be varied *ad lib*. with 'The numbers sequence might end somewhere', 'There might be more than 5 regular solids' – but all that is sheer nonsense. And the argument fails precisely because there is nothing contingent about numbers and figures. The whole point of bringing in mathematics was to show on what Moore's argument hinges – on the contingency of ordinary empirical facts. Take away contingency, and you take away all force from the argument.

This brings me to another point. Suppose that I am to solve the cubic equation $x^3 - 7x + 6 = 0$. Algebra tells me that a root exists, a number which, when put in the place of x, satisfies the equation. This might be expressed by putting the existential operator in front of the equation, like this:

$$(\exists x).\ x^3 - 7x + 6 = 0.$$

Well and good. But suppose now that I have stumbled on one of the roots, say $x = 2$. What purpose could be served by writing in an analogous fashion

$$(\exists x).\ x = 2?$$

Is this to assert that there is a number that satisfies the equation $x = 2$? But the number is already written down. Or is this to assert the existence of the number 2? And what can *that* mean, pray? If someone were to tell me, in a triumphant tone of voice, that he had just convinced himself of the existence of the number 2, I would lift my eyebrows – what the devil does he mean, and what the devil did he doubt? On the other hand, if you were assured, say, by a clairvoyant that there exists an even number which defeats Goldbach's conjecture, but a number so unspeakably large that no mathematican, not even the greatest, has the least idea of how to lay hold of it, you would feel queer. Exists! In what sense may such a number be said to exist? Of course you may seek; but you may seek until you are blue in the face; nor is there any guarantee that a method of finding will ever be found. This situation has given rise to the well-known objections of Brouwer and his school to such 'wild', unrestricted existential statements, and the attempt to purge mathematics from them.

It would appear from this that the notion of existence degenerates in two opposite directions: once when the object is *too far away*, possibly for ever out of reach, and again when it is *too near*, stares us in the face. In the latter case, the existential statement fails to act upon us as an incentive to look for the thing. It is only in the middle ground between the extremes that it makes perfectly good and proper sense, and where its legitimacy cannot be disputed.

Towards the end of his paper Moore makes a point of saying that, in a certain class of cases, 'This exists' is always true. "My view", he says, "involves, I am bound to admit, the curious consequence that 'this exists', when used in this way (e.g. referring to an after-image I am having with my eyes shut) is always true; and 'this does not exist' always false"[3]. Very curious: but it should by now have become clear where the shoe pinches. 'This exists' is indeed true whenever I utter these words – either pointing at some object or other, or, as in the case of an after-image, performing its mental equivalent, namely, paying heed to it. For everything hinges in the end on the manner in which I am pointing. Was I pointing, or was I not? If the object of my pointing did not exist, my pointing could not have been done properly – it *was* not pointing, but mere pretence. But if I point 'in earnest' and not just for the fun of it, and moreover if I am not in *delirium tremens*, in the presence of a mirage, under the effect of hypnosis etc., in short, if what I do is done in normal circumstances, then the existence of what I am pointing at is already included in the idea of pointing. (And even in the case of *delirium tremens*, or a mirage, or hypnosis, there is a sense – Moore's sense – in which *something* may be said to 'exist'.) Hence every time I say 'This → exists' I am speaking the truth; more accurately, my use of the word 'this' together with the act of pointing *creates* a situation in which what I am saying cannot be untrue: Q.E.D. Yet it is not analytic. It would be analytic if I made a gesture of pointing and said, 'Provided that I am pointing now in the *proper sense* of the word, *this* – at which I am pointing – exists', i.e. if my statement contained a reference to the act I am performing; and similarly in the case of an after-image. But 'This exists' in itself contains no such reference; therefore it would be wrong to say that it is analytic, i.e. true on purely logical grounds. Two seemingly conflicting features combine in the case. What is being said, *cannot* be untrue, at least when certain conditions of utterance are heeded, and yet it is not the sort of thing we call a *necessary* truth. But this is no more of a miracle than the fact that the statement

'These words (namely the words I am uttering now) form a sentence' is always true, i.e. each time I utter them, and yet not necessarily true. Indeed, both cases are instances of self-confirming statements, just as their opposites, 'This does not exist', 'These words do not form a sentence' are self-defeating ones.

Glancing back over the steps we have been taking to arrive at a clearer understanding, four points have emerged. There are good reasons for saying (a) that 'This → exists' comes very close to being tautologous; (b) that it comes close to being absurd, or at least, that it has no use; (c) that it *has* a use, though a rather peculiar one – namely as a clench; (d) that it is included among self-confirming statements. All these accounts sound in a way right. They dispel *some* of the uneasiness we feel. But not all of it. For the statement 'This → exists' feels disquieting in a way in which 'This sentence consists of six words' does not. There is something very peculiar about it, something not to be found in any of the other paradigms we have likened it to – neither in a commonplace like 'War is war', nor in a truth of logic like $p \vee \sim p$, nor in a bit of downright nonsense. So it looks as if something vital is still eluding us.

It can be seen, I think, that the trouble springs from the odd way in which the notion of existence is applied here. For it is *mis*applied. What I mean is this. An existential statement is a statement *faute de mieux*. As a rule, such statements are made only if more precise information is not available. Thus I may say 'There is a fox somewhere in the wood', 'there are pikes in the pond, lions in the desert', or again, 'there is somewhere an article on the use of oracles in symbolic logic, only I can't remember where'. Once I have hunted down the fox, got the pike on the line, run into the lion, spotted the article, there is no longer any need, nor any point, in saying that it exists – unless, of course, I am availing myself of the case in hand to warrant an existential statement previously made by me. But if so, the 'exists' carries reference to that statement – it makes sense only if seen in its proportion to the whole context. Generally speaking, an existential statement asserts that something of such-and-such description exists without betraying where it is. It is like saying, 'There is a buried treasure, but I am not going to tell you where'. In talking of existence we should be well aware of how near an existential statement comes to a request, or an invitation to look for something, or, again, tọ a promise. I do not mean to say that such a statement *is* a request, an invitation, a promise, only that in some respects it may act like one of these – if you scour the Libyan plain you will bump into a lion, if you fish you will catch, if you run through the right sort of journals you will come across that article, if you dig and dig, leaving no stone unturned, you will be rewarded by finding the treasure. But what can it mean to assert existence of something that stares me in the face? There it is, and that is the end of it. There is no longer any need for me to go in search of it. I have started with the end.

The general lesson to be learnt from it is this. An existential statement is related to the actual producing of a specimen of the sort in much the same way as travelling is related to arriving, seeking to finding, angling to catching, posing a problem to solving it, making an attempt to do so-and-so to succeeding in it. All these examples illustrate a certain duality of terms. This can be expressed by saying that, though each word of the one category answers to a word of the other, yet the logic of the two is decidedly different. If you are already there in my home there would be no point in my inviting you to come to my place, if I put a birthday present in your hand, I cannot promise to give it to you, if I have just found the lost key, I cannot set out looking for it. All that is a misuse of language, and nothing more. Likewise, to say of a thing that is in front of my nose, that it exists, is to attempt to speak of the one category in terms suited to the other. To caricature it, it is like describing a journey by saying 'It consisted only in my arrival', or like saying 'I approached the problem by solving it'. It is this glaring inappropriateness, I submit, that makes the act of pointing at an object and uttering the words 'This exists' so incongruous. And if I had time, I would go on to apply this lesson to Descartes' *cogito ergo sum* – which in some respects is parallel to our case, but, owing to the level-ambiguity of the words 'I' and 'exist', raises difficulties of its own.

NOTES

[1] *Aristotelian Society, Supplementary Volume* **XV** (1936), 186.
[2] This example, though not all of the implications, is due to Prof. J. L. Austin.
[3] *Loc. cit.*, p. 188.

CHAPTER XI

A REMARK ON EXPERIENCE*

We all talk of reality and experience. If we were asked what experience is or what the word 'experience' means, we should say, 'Seeing, hearing, tasting, and smelling'; saying which we may have the feeling that there is something, in common to all these cases, and that it is just in virtue of this, the common element, that they are called experience – a feature apparently lacking in anything outside this domain.

It is with the word 'experience' as it is with the word 'time': it has a trivial sense as when I say, 'This does not agree with experience' or, 'Experience teachs that metals expand with heat', or again when I refer to someone as 'a man of experience'. It is less trivial if uttered in a peculiar tone of voice – as when I speak of a 'strange' or 'uncanny experience' of mine. If the word falls into the hands of a philosopher, however, it is apt to take on a mystifying air. Has the concept of experience any definite limits, he is likely to ask, and if so, what are they? In everyday life we would not pay much attention to such scruples nor rack our brains unduly for framing a definition: just as in the case of the word 'time', we are quite happy with being able to use it in practice and don't probe any farther. But if confronted with the philosopher's question, what are we to reply? Suppose that, with a view to explaining the term, one were to enumerate kinds of experience, e.g., by saying, 'Experience is what you see, what you hear, what you smell', then the question arises: is that *all*? And further, supposing you had an experience of a kind utterly different from anything you have ever met before, would not you none the less know that it is an experience? If you answer in the affirmative, saying perhaps, 'Of course, I am quite aware that this is an experience', how *can* you be aware of it? Are not you, by making some such reply, admitting that you must already be in possession of a general idea of what is called 'experience', since, if you were not, you could not possibly recognize any new item of it as *experience*? Or is it possible that there should be doubt, that anyone should not be altogether sure whether something is, or is not, an experience?

* Composed in English, apparently in the 1950's.

The question may be put in this way: Are we in possession of a general concept of experience which remains the same irrespective of *what* we experience, or is it rather so that with any new sort of experience that comes our way the concept also undergoes a change in that its field of application is being widened? In short, has the word 'experience' a constant sense or a variable one? Before attempting to answer the question another point should be noticed. Comparing our concept with any other one of a current type,.say, that of a tree, it is easily seen that the latter one is not so sharply bounded, at least not for the layman. The question, 'Is this a tree?' cannot always be answered with certainty because of the existence of borderline cases. Indeed, nothing is easier than to imagine a situation in which one would say doubtfully, 'I'm not sure, is this a low sort of tree or rather a shrub?' Now can we conceive of a parallel situation in the case of experience? Can I imagine circumstances in which I would ask myself uncertainly, 'Well now, is this an experience, or isn't it?' (What concerns us here is not the doubt whether I really had an experience of that sort or merely dreamt I had but whether I can be sure that something or other is an experience.) If it is a question of whether there are limiting cases, things hovering uneasily on the brink between experience and non-experience, everyone, I suppose, would answer in the negative. 'Experience', it would appear from this, is a concept with well-defined boundaries. What strikes one at the same time, however, is that if anyone were to draw these boundaries there is apparently nothing left against which the concept is marked off. That is very queer – or perhaps not quite so? Indeed, is there anything with which the concept *could* be contrasted? Even if we think of a new sort of feeling, such as a new organic sensation, we should still have no hesitation in calling it an experience. It now looks as if the idea, in a sense, was an all-embracing one, resembling in this respect that of the world. In explaining the latter one, one is inclined to make a sweeping gesture, say, describe a wide circle with the hand, with the words, 'Look, all this, including space, is the world.' As with the concept of the world, so with that of experience. The impression is that it must be so wide as to cover any possible experience, even such a one as we have, in fact, never had before. And as for the boundaries, they seem to be determined by saying, 'The limits of my experience are the limits of my world'. That is, so far as I penetrate into the world – or, what comes here to the same, into life – so far reach the boundaries of experience. And when we say this, what is before our minds is perhaps a picture of a (sort of) space into which we are expanding, carrying the boundaries with us. In this conception, the word 'experience' seems to have a variable sense, adjusting itself to the wealth of sensations, feelings, etc. we have met with at any given moment of life.

Related to this is a question often ventilated – can there be a new kind of experience? For example, a sixth sense? To reply to this in the right sort of way is not too easy. What one is inclined to say is perhaps something like this. It does not seem to lie entirely outside the bounds of the possible to imagine such a thing – not of course accurately and in full detail, yet none the less in *some* sense of the word. 'I've got such a queer sort of pain', I might say, 'no, I can't quite describe what it feels like, only that it's utterly unlike anything I've ever felt before'. Well, in the sense in which this is understandable it is also understandable to speak of a new experience – although any description of it will have to be a rather hazy one, for lack of the appropriate words.

To illustrate this point, consider, for example, the case of Galton who mentions a curious faculty of his own – that he could see, in dreams only, a sphere from all sides at the same time, 'centripetally' as he says. This, I take it, will to most people be a new experience. But would they *understand* what has just been described? And would they be able to form a mental picture of such an experience? There seems to be a difficulty about this point. We have only Galton's description, i.e. the words; what we don't quite see is how to proceed from the words to the experience, or to an image of it. Yet most people would say, and feel pretty sure of it, that they do understand the sort of experience described. How is this possible? If they merely fancy that they understand, cannot they be mistaken? Or by what token, if any, can they make sure that they do? Is the criterion of understanding, e.g., the faculty of calling up an image of this sort – of embracing a whole sphere in a single perception, seeing at the same time all round it? If so, how could anyone even *try* to evoke such an image if he did not know beforehand what sort of image it was? Understanding is a disposition; but of *what?* Plainly not of calling up such an image at will, for this, admittedly, is not in our power; nor of recognizing it in case it did ever rise before our mind. (For how would you do *that?*)

The point is that there are *degrees* of understanding. In our case it doesn't seem to be a very precise one. Thus I can ask myself all sorts of questions I am unable to answer: Where exactly is the sphere located I am supposed to 'see'? 'In me?' Within the head? Just in front of the eye? In an indefinable sort of place? Has it any colour? Does it appear small or rather large? Blurred or clear? Darkish, brilliant, transparent? Yet if anyone was possessed by this curious faculty, he would understand what has been said in a fuller sense, enlarging it with all those small and yet significant details which distinguish an eye-witness account from hearsay. All the same, my inability of visualizing to the full does not seem to debar me from understanding. Why not?

What we seem to understand is at least *which way* we have to bend our effort to imagine such a thing even if we fail in the end; and besides, in making an attempt we seem somehow to be aware of whether we come any closer to the goal; and this, I think, is enough to make us say that we understand what Galton reports. In this sense, it would seem, we are indeed able to peer a little beyond the range of our actual experience, to form at least a foggy notion of what is still strange and unknown. 'Coloured hearing', the experience of the *déjà vu*, or the blind man's reputed power of seeing with the face may serve as further examples. Looking upon them, we can hardly escape the impression that experience is not like an area sealed off by some Iron Curtain but rather elastic, yielding to the power of imagination.

On the other hand, no one can imagine the experience of living in a four-dimensional space. By this I mean the 'feel' of walking about in such a world and doing all sorts of things – rowing, swimming, fishing, dancing, listening to music coming 'from the fourth dimension', or simply crossing one's hands, putting them into one's pocket, pouring wine into a glass, and lots of similar things. What mathematicians *can* describe is how a four-dimensional space would appear to an ordinary (three-dimensional) observer, but not what it would feel like to *be* a four-dimensional observer; and that it precisely the difficulty. Here, it seems, we come up against a barrier that is as precise as it is insurmountable.

The question we are concerned with could, however, be discussed in an entirely different way. Let us look upon experience, someone might suggest, with the boundaries it at present has, then it is as if certain definite limits were drawn and as if experience *filled* these limits. And this idea is in full agreement with the way in which one thinks of experience or explains the term. A popular way of doing this is to say, 'Experience is everything you meet with, enjoy, or suffer'. In this explanation the verbs mentioned seem at first to draw a sort of frame which is then filled by experience, and –

thanks to the term 'everything' – filled completely.

What this discussion shows is that an expression of generality, such as 'everything', 'whatever', occurs as an essential part of any explanation of the term 'experience'. Even if we explain the word to someone by giving him only a series of examples and he thereupon himself uses a generic term as a token of his understanding, *we* should add to our list the words, 'and in general whatever you see, hear, and feel', the word 'feel' pointing as it were to a sort of reserve space for anything that may yet turn up in future. That we are inclined to use here the word 'feel' and not any of the other ones is not without significance – hinting at the direction from which a new experience may be expected.

So there are two views – that experience is sharply bounded, and that its edge is blurred. The first of these may well be an illusion fostered by the loose way in which the above verbs are being used.

To return to the question raised at the beginning, it is a most remarkable thing, and not as fully appreciated by theorists as it deserves, that a lot of disparate cases are subsumed under the rubric 'experience'. Now is there anything each and all of them have in common? If anyone says 'Yes', we may challenge him by asking, 'What, in your view, has the experience of seeing a flower in common with a feeling of sickness?' This is a sort of question that has now and then worried philosophers. In attempting to answer it it was tempting to assimilate all these cases by reducing them to a common denominator, and this common denominator was supposed to lie in what is called 'subjective' – as if the seeing were a sensation in the eye. If this version or a similar one is accepted, it begins to look as if there really was something common to these cases, namely that they are all episodes or incidents occurring 'in' a subject. It is here that we come face to face with one of those big and subtle distortions wrought by idealistic philosophy, and by now so deep-rooted that it has become exceedingly hard to regain an unbiased outlook. The distortion consists, briefly, in this: the experience of seeing is turned into something else, namely, a feeling in the eye, or the impression made on it by an image on the retina or some such thing, so that one can say with a show of plausibility that this experience, like any other, is subjective. It is as if one said: Ah, the experience of seeing a flower is, after all, no more than a process in the eye. What is one to say to this? Well, so far as seeing is a process in the eye it is indeed subjective (as in the case that I am being dazzled by a headlight), but so far as it is concerned with things outside in space it is not.

What is wrong with this whole way of looking at things is that it creates and fosters the illusion as if a visual experience were 'in us', say, in the shape of a little image on the retina. According to modern physiology, vision is a very intricate series of processes: there is first the absorption of light by the receptor cells of the retina; this triggers certain chemical processes leading to the production of nerve impulses – a rapid barrage of sharp rises and drops in electric potential travelling along the fibres of the optic nerve; and these finally elicit electric responses in the visual areas of the brain surface. Such, roughly, is the story told by experimental research. If it is correct, and vision really consists in nothing but electric events in some manner 'felt' by us, one would expect, not unnaturally so, that what is seen should be located where these events occur, that is in the interior of the eye or the brain. Oddly enough, this consequence has indeed been drawn by philosophers no less than by physiologists. The story we are asked to accept runs about as follows. As a flower is seen outside in space and not 'against the eye', 'in the brain', or 'under the skin', this must be due to some displacement of the sensation from its 'original seat' inside the body to the external world. In other words, the assumption is being made that the little picture on the retina, or what is going on in the brain, is 'projected' on to outer space so that, instead of being seen in the eye, it 'appears over there'. This has even been hailed as quite an achievement of the human species, owing to aeons of evolution. In fact, it is a complete travesty of the truth. Though, of course, it is true to say that what a person sees depends on the events in his eye, it does not follow that what he sees *is* the events in his eye. What is seen, heard, smelt is not the *act* of seeing, hearing, or smelling, if this term is taken to refer to what is going on in the sense organs etc., any more than what is eaten is the act of eating. Yet it is this elementary blunder which underlies in part the idealistic way of thinking. And this blunder, in its turn, is prompted by the wish to equate the experience of seeing a flower to a 'subjective state' – such as the impression received by the eye.

This view, if one can call it so, is such a fantastic jumble of confusions that it is hardly necessary to dwell upon it, save perhaps by touching upon a question which presents itself naturally enough at this juncture. If the idealistic story is so hopelessly bungled, how in the first place could it ever have arisen, and further how could it ever have gained such a hold over so many eminent thinkers? (A philosopher has now and then to act the part of a detective in that he has not only to make sure of a crime also to find out who is reponsible for it. What follows is a bit of detective work.)

There are several things to be said about that. First, this whole attitude of mind becomes only intelligible if one recalls the many attempts of theorists, engaged in comparing different types of experience, to find out what they had in common; unable, however, to discover such an ingredient, they were only too eager to seize upon anything that appeared to hold out the promise of an answer, and so they hit upon the 'subjectivity' of experience. Once this point of view was reached, it seemed only natural to see in a visual experience a mode in which a subject is being affected. For to compare the fact of there being a flower with a feeling of sickness and to inquire what *these* have in common, is a thing that would not enter anyone's head. It is only through the trick of transmogrifying the sight of a flower into a subjective state or condition that these cases begin to take on a semblance of likeness.

Next, and this is perhaps more important, attention must be drawn to a change in the whole climate of opinion and the rise of a new way of viewing the world which swept over Europe at about the time of the Reformation and gathered momentum in the following centuries, but made its appearance first in England. And here we meet with one of those curious phenomena, important and yet so elusive to the student of ideas – a shifting of the meaning of a great many common words. The emergence of new words, or a change of the meaning of old ones is never a phenomenon to be dismissed slightly: it portends something going on at a great depth. What was actually happening is very difficult, if not impossible, to say – as no one really seems to know what produced these changes, and, moreover, as there are good reasons to suppose that most of them came about unconsciously.

Faute de mieux we use such metaphors as 'a change in the currents of thought and feeling', 'the emergence of a new consciousness', etc. It is in this figurative sense only that they shall be used in the subsequent observations.

What strikes any observer of that time is its growing subjectivity. Now to what is this due? There seem at least to be two centres of diffusion. Montaigne, the first great self-observer in modern times, is certainly one of them – that solitary figure that was so far ahead of his time as to be, intellectually, much nearer to the thinkers of the eighteenth century, especially Hume. He spoke, for instance, of the 'flux of ideas' in his mind. This trend was, despite all differences, further helped by the philosophy of Descartes and, perhaps even more so, by the writings of Pascal, the first one to speak of 'the logic of the heart'. Another source must, no doubt, be seen in the great movement of the Reformation, with its insistence on the 'inwardness' of matters of religion. All this brought about a subtle change in men's feelings and their attitudes which finally left their deposits in English – in part by giving new meanings to old words, or borrowing them to express some newly-awakened but still half-conscious feelings, and in part by creating fresh words and verbal patterns.

To enter freely into the minds of those late medievalists, lacking as they were in so many of the ideas we could not do without today, to recapture the atmosphere in which they were living, is denied to us. But by tracing the history of certain key-words we can, to some extent perhaps, recreate for ourselves the spirit of that age and thus attain a heightened awareness of some of the major changes of human thought. For the history of an idea is written in the history of the word embodying it.

Our word 'interesting' is itself an interesting instance of that change. Like a great many other words to be discussed presently it was unknown to the speakers of English in the fifteenth, sixteenth, and seventeenth century; it turned up only in the next one when it first meant 'important', and it acquired its present meaning in Sterne's *Sentimental Journey* (1768). About the same time the verb 'to bore' cropped up. If we really wish to enter imaginatively into the spirit of those far-off ages, we must first of all divest ourselves entirely of these modern and 'subjective' notions and try to regain an outlook where the things mattered and not the observers with their specific feelings and reactions. Another of those characteristic words which point to a change in thought and feeling, not only in England but in France and Germany as well, is 'romantic'. It occured first in this country in 1659, but its sense was then very different from what it means today: it simply meant 'like the old romances', with their fabulous horses, dragons, and magicians, or with their turgid and inflated sentiments – things which were utterly discredited when judged by the standards of the new Age of Reason – so that the word was first used as a term of depreciation, associated with ideas of the unnatural, ridiculous, bombastic, and childish. Early in the eighteenth century, however, with a gradual shift of feeling towards wild, untamed nature, the word began to absorb into its meaning fresh values, above all the sense of 'enchanting to the imagination' – as when we speak of a 'romantic adventure'. Besides that, another sense came to the fore which is still strongly felt in the word: it came to express the growing sensibility to wild, lonely, or fantastic scenes, reminiscent of those earlier tales, and conse-

quently to be applied to them – to old castles, moonlight, and moors; until
its sense shifted still further to denote 'the touching impression' we receive
from them. Like 'interesting', or like 'picturesque', which came by way of
France from Italy and cropped up at the same time, 'romantic' is one of
those modern words which register, not so much the external things and
their qualities, as the response they evoke in us, slightly tinged with emotion, and throwing into relief the subjective side of an idea. And when we
look more closely at the subjective feeling indicated by 'romantic', we find
that it is of a literary kind: it refers to nature seen through the coloured
glasses of fiction and poetry, just as 'picturesque' meant 'fit to be the subject
of a captivating picture', thus referring to nature as seen through another
glass, that of painting.

The adventures of these two or three words are typical of the age, shared
as they are by many others, and it may be useful to compile a list of some of
the more important of them, together with the dates of their appearance,
according to the *Oxford English Dictionary*.

despairing (1591 Shaks.)	enravishing (1681)
moving (1591 Shaks.)	sympathetic (1684)
tempting (1596 Shaks.)	refreshing (1697)
touching (1601 Shaks.)	entertaining (1697 Collier)
attractive (1602)	shocking (1704)
'taking' (1605 B. Jonson)	amusing (1712 Addison)
enchanting (1606 Shaks.)	entreating (1718 Rowe)
dissatisfactory (1610)	affecting (1720 Rowe)
enthralling (1640)	depressive (1727)
exhilarating (1643 Milton)	dispiriting (1733)
fascinating (1648)	pathetic (1737 Pope)
diverting (1651 Baxter)	bewitching (1749)
catching (1654 Baxter)	striking (1752)
transporting (1655 Sidney)	disgusting (1754)
charming (1663)	thrilling (1761)
surprising (1663)	interesting (1768 Sterne)
enlivening (1664 H. More)	sentimental (1768 Sterne)
endearing (1667 Milton)	impressive (1775)
engaging (1673)	repugnant (1777 Watson)
captivating (1675 H. More)	depressing (1789)

repulsive (1792)   revolting (1806)
repellent (1797)   exciting (1811)

All these are words which describe things and situations not objectively, as they are, but by the *effects* they produce on us. As shown by the list, it was a general shift of the mental atmosphere which found expression in a change of the meanings of many common words, the addition of fresh senses of values to existing ones, and in the coinage of new terms; a shift that began roughly with the seventeenth century but which, for the most part, went on in the next, the century of Berkeley. It goes without saying that most of these semantic changes were unplanned and rather due to a half- or unconscious groping for ways of expressing the new modes of thought and feeling which began to rise all over Europe at that time, the 'age of ferment' as it came to be called. It was a movement too fluid and flickering to be traced with precision. Only its broad outlines are clear. If I had to compress into a few words what it all amounted to, I would say that underlying all these changes in detail there was a general and very real shifting towards the subjective in our experience, a trend which was transforming the whole mental scene.

I must apologize for intruding into a field where I can feel to be only an outsider; nor do I think that what I have to say is more than a 'hunch', to be confirmed or rebutted by more extensive research. The hypothesis I submit is (1) that there were far-reaching semantic changes of a very similar kind in English, French, German, and the Latin as it was written at the time; (2) that all this points to a sharpened awareness of the subjective element in experience and, what goes with it, to a shift of language towards describing (imagining, evoking) things from the observer's point of view; (3) and that it was this new and greatly heightened awareness which has done more than anything else to pave the way for, and later sustain that particular way of looking at the world which goes under the name of 'idealism'; and by this I mean it was those strong currents of thought and feeling – as reflected in the changes in these European languages – which made such a view acceptable and lent it even a measure of plausibility.

If there is any truth in what I am saying it goes a far way to explain why idealism appeared on the scene just at that particular time, with its growing sentimentality, an age not yet hardened, plastic, groping for self-expression. At other times, e.g. during the renaissance, it may by no means have felt so natural and convincing: early in the eighteenth century, however, it became presentable, owing, no doubt, to the greatly changed atmosphere of the time; and as this was particularly open to the subjective interpretation of our sense experience, etc., it must have appealed very strongly to men of the stature of Berkeley or Hume. Thoughts may lie dormant for a long time until they find a more favourable climate, and then they break through to a dominant position. Thus the subjectivity of the sense qualities (warm and cold, for instance) was already known to Democritus but had found no resonance in his days – there is no trace of it in Plato or Aristotle – whereas its rediscovery by Galileo, Boyle, and Locke did have mighty reverberations, coming at a time when Western Europe was prepared, indeed famishing for just such a thing. An idea like this, one might say, acts as a centre of attraction for other more dimly perceived ones which precipitate around it like tiny ice-needles in a cloud; and these ideas, most finely distributed in the air, yet not too clear and little more than vague impressions, come gradually to form the dominant tone of thought of an age. But this is perhaps too strong an expression for what may better be described as a mere instinctive attitude of mind, one thoroughly alive to certain ways of thinking, feeling, and interpreting the world while quite refractory to others, an attitude as yet unformulated – until it finds expression in a philosophy. That is one of the services done by philosophy: it co-ordinates the 'living' ideas of a time and integrates them into a certain pattern. The metaphysical systems of a time often bring out this dominant pattern, 'the figure in the carpet'. (Hence their curious family likeness). It is a process like crystallisation, to borrow Stendhal's image: just as a branch thrown into the depth of a shaft in a salt mine is later found covered all over with brilliant crystallisations, in a similar way the floating and fluid ideas of a time may undergo crystallisation, and then, almost suddenly, a new way of looking at the world comes into existence.

I have spoken so far as if crystallisation were a spontaneous, almost mechanical process. But the paradox is that, unlike customs, changes of language, and lots of other things, it can take place only in a single mind, a philosopher (not necessarily a professional one), and if it does, he becomes the voice of his time. There is no such thing as an anonymous philosophy, apart perhaps from the early speculations of the East. The particular shape given to a philosophy is of less consequence, compared to its one important quality: to give expression to the major trends of the time. And provided this is

achieved by a philosophy, it comes to be a symbolisation of the *Lebensge-*
fühl of a whole epoch, in a similar way as a style of art, say, Baroque mir-
rors the mood of a time. But it would be a fateful error to assume that such a
philosophy owes its authority to *logical reasons.*

The philosopher, though not a magician and certainly not the creator of
all the ideas budding in his age, is rather their architect: harmonising the
many recalcitrant elements, fitting the broken fragments of vision together
so that the world takes on a new depth and a new significance, and in all
that catching the spirit of the time. In order to do this he must, to a higher
degree then anyone else, be alive to what is in the air, or – at times – even
have an anticipatory glimpse of what is coming. That is why there is so often something of the seer about him. I do not mean to say that a philosopher cannot have greatness and yet be a solitary figure – like Democritus,
Montaigne, and others. However, it is not my intention to discuss the question in general here, but to confine myself to the case of idealism.

In this case, it was the rising tide of subjectivity that swept it into domi-
nance – a trend recognizable on many planes, the linguistic, religious,
moral, literary, and the personal. It is against this background that the as-
cent of that philosophy must be seen. Indeed, it cannot properly be understood unless it is viewed in this light, namely, as the reflex of the growing
sense of subjectivity. This, and not the arguments put forward by their pro-
tagonists, gave it its tremendous power and verve. Generally once one
comes to see the background of a philosophy one is no longer so interested
in the 'grounds' proposed for it.

It is a curious spectacle, this rise of idealism, letting us catch a glimpse of
what stands 'behind' a given philosophy, what sort of forces carry it up to a
dominant position and, on ebbing, leave it stranded again. The mood of a
time may be propitious to Berkeley, and then swing to Kant or Schopenhauer, or to Kierkegaard or Heidegger; and each time this particular philosophy is regarded as the Last Word. And in a way it *is.* (Much could be
said about this 'atmospherical element' in Leibniz or Kant.)

But what about true and false? I would suggest that the philosopher
should not let himself be unduly disturbed by that. To ask whether some
metaphysical vision of the world is right or wrong is almost like asking
whether, e.g., Gothic art is true or false. What *can* be asked – and I am not
even quite sure of that – is perhaps whether a certain work of that style 'fits'
into a given scenery, surrounding, or a time; and much the same, I think,

applies to idealism: i.e. it can be argued that it was an outlook 'fitting' at the time it made its appearance but no longer so today. True, a philosophy is made of thoughts, and arguments are used to support it. But that they can prove anything is a myth which is just disappearing. If philosophy were a domain of specialised knowledge, it would be proper to speak of its truth or falsity. In contrast to science, however, the life of philosophy consists much more in the breaking down of barriers, the changing of the atmosphere, the developing of a sharpened awareness of aspects of thought or feeling, the seeing of new patterns; and such achievements cannot be measured by the standards of true and false, not wholly so. Nor do I think that the question of consistency, that battle-ground of shadow-boxing, is really of so much relevance as is made of. 'Was Berkeley consistent in his account of corporeal substance?' As if that mattered; as if I could not take any number of assumptions as come into my head, with or without meaning, and derive their consequences, with as much rigour as I can: yet what would be the point of it? It should be a sort of exercise in logic, yes – but otherwise devoid of everything that is the life-blood of philosophy. And supposing that the improbable should happen, that a philosopher should ever turn up who is wholly consistent, would it be due to this that his work would have a claim to significance or grandeur? If not, why this insistence on logical flawlessness? And whence the belief that you can 'refute' a philosophy by digging out some internal inconsistencies? I take this to be a relic of scholasticism; and I am provoked to say that what really matters is, for the most part, what is between the lines; and this, of course, eludes such analysis. I would even go further and say that a philosopher may write a book every sentence of which is, literally, nonsense, and which none the less may lead up to a new or a great vision. (Schopenhauer is nearly such a case.) If I am right in this, it shows why a philosophy, if it springs from an authentic vision, cannot be destroyed by argument, not even by contradiction. So what I have to say reduces itself to two things: (a) if a philosophy is consistent it does not mean a thing; (b) if it is inconsistent, since you can usually wriggle out of it by the delightfully simple trick of reinterpreting some expressions, it does not mean a thing. But this, I am afraid, will be lost upon those who are using live ammunition in shooting at idealism, or else defending it – instead of understanding it in its historical aspects.

Indeed, instead of taking sides, the modern philosopher should ask himself, 'How is it possible that men of genius like Berkeley and Hume could get so lost and embrangled as to swallow the idealist story?', a question on which curiously little light is shed by the post-mortems of the customary kind. Yet the fact remains that they not only swallowed it but took it as a matter of course, as something so evident and only in need of being pointed out as to win at once general consent. This at least is the impression one gets from reading Berkeley. There must certainly have been something 'in the air' of that time that made such an outlook not only acceptable but to all appearance cogent, unavoidable.

CHAPTER XII

THE LINGUISTIC TECHNIQUE*

"We are not analysing a phenomenon (e.g. thought) but a concept (e.g. that of thinking), and therefore the use of a word."[1] This, in a nutshell, is the linguistic approach. My objection to this is: (1) In order to see how a word is used, and above all whether its use in a certain concrete case is 'correct', 'right', etc., we have to visualize at least *some* situations in which it would be proper to use it, and in order to do that we have to pay heed to the phenomenon (e.g. thought). The two things cannot be separated. (2) Analysing the use of words, important as it is, does not always get us far enough. The use of language may now and then, not too often though, contain hints of something that does not come out quite openly – especially when it is a matter of idiomatic use, arbitrary, capricious, utterly incalculable as it is; yet it *points* to something: but to point at is not to establish a thing, subject or point. It is for the philosopher to get at the *rationale* he may, rightly or wrongly, feel to be behind the use. Modern logicians, for example, were given a tremendous fillip when catching a glimpse of a clear and rational pattern that seemed to underlie the use of such words as 'not', 'and', 'or', 'if', etc. – extracting from it the idea of a truth-function. This was a constructive step going far beyond recording the use or analysing it. In connection with this, mention should also be made of such happy findings as the discrimination between 'task' and 'achievement words' (Aristotle, Ryle) which has done so much to dispose of philosophical headaches. This, too, means to penetrate to an underlying idea which makes us see the reason why the use is at it is, making certain features of it transparent to our understanding. This rationale is never quite on the surface: it is down so deep that it needs an uncommonly perceptive mind to descry it and dig it up so that it may be seen by everyone. That's why it is possible to make discoveries in philosophy, a thing denied to the recorder or lexicographer of language. (3) Assuming even that a certain word, say, 'certain' and 'true', or 'believe' and 'know', is in fact used in such-and-such a way, this need not *bind* us to this

---

* Composed in English, evidently after the publication of L. Wittgenstein's *Philosophical Investigations* in 1953.

usage. For if we look into it more closely we may come to find very considerable differences beneath the surface, differences which may make us dissatisfied with language as it is, perhaps so much so that we may prefer another way of using it. Just noting how language *is* being used and leaving it at that is not enough. The linguistic approach rests, after all, on the assumption that ordinary language is *adequate* – an assumption I am greatly inclined to challenge. But to see whether language is inadequate, and if so, where the shoe pinches, one has to look at the facts themselves. Let me remind you of a few quite obvious cases.

It has been said that intentions may often be left half- or even quite unformulated, and that putting them into words makes them at the same time definite – a very true and penetrating remark, yet not one gained from paying heed to actual usage. Taking a broader view, one may perhaps say with Sartre that "in speaking we always say more than we intend to". (By the way, precisely the point which makes it so difficult in philosophy to say no more than one knows.) It is as if words had a life of their own, and one most difficult to bring entirely under our control. This is noticeable to anyone who cares to look at idiomatic speech, "the life and spirit of language" as Landor said – with its idiosyncrasies, its eccentric, at times even illogical constructions, its expressive phrases the meaning of which could never be guessed from that of the words which compose it. Why, for instance, it is a 'grass widow' in English and a '*Strohwitwe*' in German is, so far as I know, still obscure to the experts. However, it is not this feature of idiom, curious as it is, upon which I want to enlarge. There is another and more sinister aspect of the matter. As Logan Pearsall Smith observed, the

expressiveness of irrelevant phrases... seems to show that there is a certain irrelevance in the human mind, a certain love for the illogical and absurd, a reluctance to submit itself to reason, which breaks loose now and then, and finds expression for itself in idiomatic speech.

But how does this square with what has been said before on the existence of some rational principle beneath the surface features? Perhaps it does not; and if not, there seem to be two opposing tendencies in human speech, one suggesting the operation of something rational, the other a rebellious, impish, will-o'-the-wisp spirit (the fascination of the absurd). After all, language is made in the image of man, and is just as inconsistent as he is. I incline to come more and more to the conviction that there is indeed a dual aspect in language, a tension between the rational and an irrational, indeed anti-logical element, that seems to take delight in flouting the laws of grammar and logic — just as there is tension among the 'reasonable' reasons and motives, the wishes, inclinations, promptings, the acquired habits, the sudden 'irrational' impulses which are apt to sweep over us, and what not, all of which affect our behaviour. To mention just one significant fact: if there ever were any human language with quite strict grammatical rules and without these 'illogical' expressions in which speech indulges, without those occasional departures from the beaten track, those sparkling little nucleuses of life, it would at once strike us as dead as a lunar landscape, – an arid desert of reason and correctness. Does this not tell a tale? But happily, there is a marriage of reason and unreason in language. The point of mentioning this will appear later.

Apropos of adequacy, take a phrase like 'the stream of consciousness'. It is easy enough to picture intellectual life flowing along like a stream. Yet, if I savour the phrase, I feel uneasy. The picture of a stream seems somehow not quite to fit: surely it is not so that everything in our mental life is in an incessant state of flux. On the contrary, while one's attention is absorbed by one thought, say, a problem or a difficulty, there is no flowing – there is just this one thing, the difficulty, looming large and filling the whole mental field. Nor do I find anything particularly 'streamy' while seeing the point of a joke, looking at a philosophical argument, and trying to pick holes in it, etc. A feast of reason is not a flow of soul. This impression, that what we experience is a flow is much more likely to arise when one has to listen to an uncommonly dull lecture, gazing abstractedly at a fly (say, or else in one of those rarer philosophical moments when one all of a sudden tries to catch the present instant so that it is arrested, spit on a spin[2] as it were and offered for inspection at leisure – and is then left empty-handed and with a long face. And as for the leaps of attention, they fit even less into the picture of a stream. W. James used a very infelicitous metaphor when he likened consciousness to a 'river' or a 'stream' with 'substantive' and 'transitive' parts – the last one embodying a notion which is itself prompted by a bad misconception of the workings of language. (Just as there is a 'feeling of *blue*' and a 'feeling of *cold*', so, he thought, there is a 'feeling of *and*' or a 'feeling of *if*', constituting the meanings of these words – only that the latter ones are too swift to be caught by simple inspection, being of a very peculiar and elusive nature.) No wonder, then, that a good deal of flim-flam has been talked about James Joyce, Mrs. Woolf in particular, and the stream-of-conscious-ness-novel in general which, so it is claimed, pictures vividly the process of thinking with its tangled fringes of association – wrongly claimed, I should think, because everything, even the inarticulate and nameless is magnified into words, thus giving a bias to the whole way of presenting conscious life. Now if the use of language were really a sort of Supreme Court as many people seem to believe, one from which there is no appeal, I should simply have to say, 'Look, the existence of that phrase, and the currency it has gained in common speech is proof enough that consciousness is in the na-ture of a stream' – a remarkably easy way of exploring reality by disregard-ing it. But, and this is my whole point, we are not *slaves* that we should have to submit meekly to the authority of the existing use. In the case in ques-tion, there is indeed a somewhat strained relation between metaphor and truth. Now in criticising the appropriateness of the phrase I do not care a pin for the actual use of English; what I am trying to do is rather to look with a fresh eye at the phenomena and see how far language fits the facts.

There is another reason why this phrase is sometimes felt to be so apt and fitting, and this is connected with the idea that everything in in a flux. Sup-pose that I am looking at a chest of drawers in front of me; nothing hap-pens, at least not in the ordinary sense, everything around me remains the same; and yet I am curiously aware that, while being all attention, some-thing is changing all the time, that even while I am looking fixedly at the chest, thought always lags one step behind as it were, my words being un-able to catch up with the experience speeding ahead of it. It is this, I believe, which made certain philosophers wish to begin philosophy with a sort of in-articulate sound – as if to point to the fleeting focus of attention that defies articulate expression. In such moments it *may* seem that experience is a stream.

I have so far been speaking of the adequacy of language which has, quite wrongly as I hope to show, been taken for granted by most of the paladins of the linguistic approach. Whether language is really adequate may fur-ther be discussed with the help of another illustration. Paradoxical as it seems, there are cases where neither the expression 'real' nor 'unreal' or 'ap-parent', neither 'existent' nor 'non-existent' is in place. The slowing-down of a clock in motion is an example. This has been hailed by some mystery-minded people as a confirmation of the age-old idea that time is 'illusory', 'subjective', 'unreal', whereas others insisted that time flows on at an even rate entirely unconcerned with the behaviour of any clocks or instruments. On the first view, the clock's retardation is 'appearance', a sort of trick played on us by nature. Nothing could be further from the truth than this. In the case of a conjurer's trick, you are 'taken in', as the saying is: if you had only watched more closely you would have noticed how it was done. Not so in our case. No matter how careful, meticulous the observers are, what sort of clocks are chosen for the experiment, and how well synchronized they are at the start, the outcome will always be the same. In this sense, there is nothing 'illusory' about the slowing-down. As for the word 'subjective' it suggests that the experiment may be depending on the observer's idiosyncrasy, taste, temperament, etc., while it could be carried out with self-recording instruments without a human agency at all, and yet with the same result. What is called 'relativity of time', is only concerned with a new and unsuspected property of clocks, namely the dependence of their rate on their state of motion. A moving clock goes more slowly than when at rest – this is the upshot of the matter.

If the retardation is not apparent, is it real? Is it, for instance, a phenomenon caused by motion? In the eyes of Lorentz and other leading physicists of the time the earth was indeed moving through the ether, but the effect of the motion, the 'ether-wind', was supposed to be cancelled by a shortening of the arms of the instrument designed to detect the motion, and a corresponding slowing-down of clocks. On this view, there are two processes going on, both of them real: (1) motion through the ether, (2) contraction of bodies in the line of motion and slowing-down of clocks. The two are related as cause and effect. If there were not a sort of occult agency meddling in the instruments, the earth's motion could easily enough be detected. As it is, the two processes are so delicately balanced as to compensate one other, that in spite of motion no trace of it is to be found. It almost looks as if nature were animated with the malicious desire to foil any attempts on the part of the scientists to chart our path in the trackless ether-sea. It is here that the power of the 'verification principle' makes itself felt. Reflecting on the matter, we see that we can only speak of a motion through the ether or a shortening and slowing-down of measuring-rods and clocks if we are in possession of a method, or if we can at least indicate one, by means of which, in any given case, it can be decided whether or not something is in a state of such a motion, and whether or not it is being shortened and slowed down. In the absence of any such procedure, we simply deceive ourselves when we think that a clear meaning can be attached to any of these statements, however strongly this may be suggested by their outward grammatical appearance. When the idea of an ether-drift is dropped as being undiscoverable in principle and not merely for technical reasons, our whole outlook is apt to change.

To return to our question, whether the slowing-down is a causal effect of motion, to say that it is presupposes that the retardation is an actual process. Suppose that, your watch being slow, you go to the watchmaker and he tells you that some grit has got into the works. In this case, something has actually happened to the watch: the grit increases friction and this retards the watch. Now does motion act on a clock in any similar way? Has the clock, perhaps, been damaged through the motion so that it fails to show the right time? Certainly not; nothing 'happens' to the moving clock to slow it down, as can be seen from the fact that the same clock may for one observer be retarded and yet for another unretarded. But if so, it is only natural to say that the clock 'seems' or 'appears' to be retarded. Well now, it may be asked, if these are 'appearances', what then is the truth behind them? What is 'really' happening? To this there is no answer; and there is none because the whole question is couched in the wrong terms. A moving clock *is* retarded, certainly, but in a relative sense only – as judged from a system of reference that may be chosen in any manner we like. And this makes the case so unlike that of the watch. The grit in it is not 'relative grit' – it is there or it is not there, and if it is there it is the cause of the retardation. But to ask in a similar way, 'Is this clock in motion or is it not?', without adding *in respect to what* it is in motion makes no sense. As there is no such thing as 'absolute motion', motion with regard to empty space (or the ether), it cannot be described as the cause of anything, and therefore not as the cause of the slowing-down of the clock. If, however, 'motion' is taken in its relative sense, then to seek the reason for a clock losing time in its motion relative to (say) another one only leads to a paradox: for, since there is complete symmetry in their relation, the second can with equal right be said to be the cause of the first one losing time. Just as the concept of motion is a relative one, so is that of retardation. A certain body is, and is not, in motion, depending entirely on the system of reference chosen. And so it is with losing time. To approach the slowing-down with the alternative 'real or apparent?' is an exercise in Procrustean art. In truth, the retardation is neither real nor unreal, neither fact nor fiction, neither existent nor non-existent. These categories, crystallized out of common speech, prove unable to cope with a situation like this.

The same account holds of a moving rod and the shortening it undergoes; and at the risk of being boring the analogy may be pursued a bit further. Somewhat as a liner is compressed between the screws driving her forward and the sea holding her back, so, it was thought by Lorentz, a material rod is being squeezed together as a result of its journey through the ether. A very uncomfortable idea – especially when one considers that the rod, on reaching a certain speed, will be completely flattened out so as to appear as 'height and breadth without length', shrunk to two dimensions, its cross section. It really looks now as if something very alarming is going to happen to a travelling rod.

Einstein, starting from two general principles established by experiment, arrived at the same formulae as Lorentz, but without introducing any *ad hoc* hypothesis. Mathematical analysis shows that, if his two assumptions are correct, moving bodies and clocks must behave *as if* they were subject to such a distorting and at the same time retarding force. However, the interpretation given to those formulae is now an entirely different one. With the notion of an ether-drift gone, there came also an end to that of a causal effect of motion as a ghost-force capable of squeezing bodies together, retarding clocks, or playing any other pranks: both, the shortening and the slowing-down are simply the outcome of the 'logic of measuring', if I may use this handy phrase.

One last word about the concept of 'real' before leaving this subject. Suppose that something is called 'real' about which all observers agree, such as that two rods are in contact. Is the shortening now real? As a moving rod will be found shortened by an observer on the ground but entirely unshortened by one travelling with it, the shortening is not real in the sense defined. Alternatively, if something is called 'real' if its existence can be established by experiment, then the contraction *is* real. However, there is little comfort to be got out from this; for the contraction of a rod of length l, being dependent on the observer's state of motion (which is quite arbitrary), will now assume *any* value between 0 and l, or *all* these values at the same time. To say of something so utterly indefinite as this that it is real runs counter to all our ideas. So the second definition is no good either. To describe therefore the shortening either as 'real' or as 'unreal' is to misdescribe it. Is language adequate?

The real antithesis underlying this whole discussion is that of 'relative' and 'absolute'. The pure 'space-distance' of two events with respect to a system, just like their pure 'time-distance', is relative. Three years after Einstein's first paper on the subject, in 1908, Minkowski made a momentous discovery. If we look upon happenings in ordinary space and time as a sort of 'points' and arrange them in a four-dimensional 'event-space', then the distance of any two events, measured in this space, has the remarkable property of being 'invariant', i.e. to be the same for all possible systems of reference. That is to say, while observers will find themselves in disagreement as to (purely) spatial and (purely) temporal distances, there is one thing about which they all agree, something neither spatial nor temporal, indeed inaccesssible to our intuition – the 'interval' between two events (into which length and duration enter, so to say, as equal partners). Moreover, the mathematical aspect of relativity reveals a complete symmetry between space and time coordinates, pointing to a sort of underlying union of both. ('What nonsense!' – a growl from the philosophers – 'time is time and space is space, and no discovery can change that'.) Thus, whereas Einstein established the relativity of length, duration, and a number of other physical concepts (such as 'simultaneity' and 'shape'), Minkowski discovered the 'absolute' world behind them all – so that the career of one aspect of modern physics may be summed up in the phrase 'from the relative to the absolute'.

Viewed from this angle of vision, the changes which measuring-rods and clocks undergo, or seem to undergo, appear now in a new light. Mathematically, the Lorentz-Einstein formulae simply describe a rotation of the coordinate system in four-dimensional space-time, so that those queer changes which rods etc. experience correspond to changes of perspective. Just as, when one looks at an object while moving about, it will appear foreshortened in certain ways, though it does not change itself, so a four-dimensional entity, e.g. an interval, does not change, while its perspectives, or better its projections on ordinary space and time do. That is, an interval can be split up in any number of ways into a 'space-like' and a 'time-like' component, each of them answering to a particular manner of measuring a space- and a time-distance.

From the vantage point thus reached it is easier to see why the old question 'real or unreal?' 'fact or non-fact?' cannot be answered. It cannot, because new categories, better adapted to the newly-discovered facts, cut here through the old ones. 'Reality' and 'appearance' are the wrong categories: nothing but confusion results from applying them to the present case, whereas 'relative' and 'absolute' are the right ones.

Anyone who takes an unbiased view of the situation will, I think, agree that we meet here with an entirely new phenomenon – the more or less conscious moulding of new categories on the part of the scientist, of new modes of thinking and describing. Take, for instance, 'absolute', in the sense of 'invariant': needless to say, it is not a 'natural' category abstracted from the forms of word language. It is rather one of those bold constructions characteristic of science in which imagination of the highest sort, matter-of-factness, and mathematical precision are curiously blended. Yet it is these notions 'relative' and 'absolute' which have done much to transform our whole world picture. Uncommon sense, not common sense, is required to achieve such things. Indeed, it can safely be assumed that Einstein or Minkowski, had they ever subscribed to the common-sense view, would never have made their discoveries: being bogged down at every step by their trust in it and, besides, being held captive in the forms of language, the very embodiment of common sense - how could they ever have broken through to a radically new way of looking at the world? Yet that is precisely what one may have to do to cut loose from worn out grooves of thought.

This calls our attention to another aspect – the dangers that are sure to arise when language is taken at its word. We are undeceived on this score only by bitter experience. Remember merely the countless analogies, so natural to language, and the puzzles they give rise to in philosophy – to that, for instance, which St. Augustine found in the measurement of time. What constitutes the breath of life of language in its naïve, unreflected use becomes a mortal danger to clear thinking. What I have in mind is not so much the figurative mode of expression as the far subtler deception inherent in the fact that concepts present themselves in the form of stable grammatical categories. A metaphor can be avoided or seen through; the grammatical categories permeate thought itself.

This suggests a subject so that it can only be slightly touched upon – the way, namely, in which the forms of language have influence over thought. Take, for example, the case that an idea is expressed in the form of a noun. This makes us instinctively look out for something that is designated by it, just as a man is by his name. How strong the propensity is to seek for an object correlated with the noun can still be seen from certain ways of thinking that now and again have dominated science. Thus a force would often be thought of as something actually existing – lurking in the dark and ready to pull like a stretched spring, until men like Kirchhoff and H. Hertz did away with it by showing how a system of mechanics can be constructed without the idea of force and its attendant difficulties. Likewise heat and energy were lent the attributes of a substance – very much, one suspects, on account of their appearance as nouns - just as in the case of the 'electric fluid' of which there was so much talk at the time of Franklin. True, there were other reasons besides that: the discovery that energy cannot be destroyed or increased, its amount being constant, did much to foster the view that behind every change there remains something that keeps identical with itself and merely alters its form. From this point of view, physics is nothing but the study of energy and its transformation into new forms. Yet the rise of this world picture, despite its many defects (and abandoned since), is only understandable when one sees how captivating it must have been to the imagination – holding out a prospect to explain all natural phenomena in terms of an underlying substance and its changing forms. And was it not ideas of a similar kind which, partly masked by others, led to the assumption that a certain 'spirit' must be present in any combustible body which escapes with the fire – the phlogiston?

To mention a less sophisticated case, it is in part the characteristic appearance of a flame, but in part also language which creates the impression that it is a 'thing' (and a 'living' one into the bargain) – considering perhaps how perfectly natural it is to say that it has to be 'fed', sends out 'tongues', 'licks up' in passing, or 'leaps up', 'consumes' a body, and 'dies'. (If we had in our language a verb instead of a noun to refer to a flame, would this way of apprehending still be so natural to us?) And further, is it really of so little consequence whether we find in our language nouns like 'soul', 'mind', 'spirit', the 'ego', 'conscience' (semipersonified as that inner monitory voice), 'genius' (tinged with associations of 'demonic possession') alongside of such words as 'hunger', 'lightning', 'thunder', 'rain', 'wind', or 'fury', 'fate', 'time', etc.? All of which suggest, though to a varying degree, an interpretation in terms of a substance, or an activating agent or principle. (See Shelley on inspiration, Coleridge on imagination, Yeats on emotion.) Even to a verb like 'to exist' there seems to cling a faint undertone of a sort of activity – as if the verbal element had modified our apprehension. (See Heidegger on the *Nichts*.) In this connection it is perhaps worth recalling that in Frege's view logic is concerned with exploring such things as 'and-ness', 'or-ness', 'if-ness' etc. – as if these were a sort of supersensible objects, yet each of them endowed with well-defined properties. Would philosophers who had been growing up in an entirely different language, with the functions of nouns far less developed, be found on the same paths of thought, or arriving at essentially the same world picture?

Considering some other languages such as Greek or German – where owing to the gender and therefore the ease with which nouns of any sort combine with 'he', 'she', 'it' – it is only natural to expect that the personifying and myth-creating power should be much more alive, or at least dormant. Indeed, where you have to refer to day and night, or the sky and the earth as 'he' and 'she', a splash of the mythical will be near at hand. So Schelling was, after all, not so wrong when he compared language to a 'faded mythology'. A word in such a language has not only a meaning but something else as well, not quite easy to describe, something that surrounds it like an atmosphere – a dim halo or aura, formed of all those countless figures of speech to which it lends itself or into which it fits. This aura, so important to the feel of the word and yet so elusive, cannot be recaptured in translation. Of this halo, the word 'sun' gives perhaps an illustration: in German *Märchen* it is always personified as an old woman, owing to its feminine gender there – a thing quite unnatural to an English-speaking person; for other reasons, '*le roi soleil*' cannot be rendered in English in such a way that the naturalness of this image, and with that its aura, is preserved. To cite another and more serious example, the masculine gender of the noun 'logos' was not the least among the factors that smoothed the way to the late Hellenistic interpretation: the Son of God. Yet this aspect of the matter, important as it is, is constantly overlooked in any analysis of 'meaning'.

A whole essay might be written on the question how certain words, taken from daily life, take on a metaphysical significance – for instance, the 'Logos', 'Reality', 'Necessity', or '*Wille*', '*Werden*', '*élan vital*', '*Angst*': terms which have a trivial use in ordinary life, but are later transcendentalized and acquire metaphysical stature. Hence that familiar tendency of theirs to hover between triviality and a new sense charged with deep and mysterious significations.

Our present concern[3], is with science. And here it must be said that besides the parts of speech – noun, verb, etc. – grammatical forms in a much broader sense have exerted a guiding as well as a restraining influence on thought. Thus it seems a truism to say that a predicate must be a predicate 'of' something. Well and good, so far as ordinary language goes; but when light is conceived of as consisting of certain undulations, does it necessarily follow that these must be undulations 'of' something? Yet so strong is the hold of grammatical moulds upon our mind that we feel driven to such a conclusion, almost irresistibly, delusive as it is. It was, I think, Lord Salisbury who defined the ether as the nominative case of the verb 'to undulate', thus bringing out the linguistic fallacy underlying the whole concept formation. Science, looked upon in this way, is to no small extent a struggle to get rid of the forms imposed on it by word language.

So far it was the influence of certain linguistic categories upon thought which claimed our interest. But mention must also be made of another sense of 'category' which came to gain a vast influence in philosophy. In this sense – to pass over Aristotle's account as of less importance for the present purposes – categories were thought of as ultimate forms of being as well as of thinking. Not being merely linguistic, they should not be confused with "a table of classification of all nameable things" as J. S. Mill has it (or in modern parlance, with 'logical types'). The underlying idea seems rather to be: concepts which are necessary, that is, without which the attainment of a coherent body of experience, or of a comprehensible world picture, would become impossible. Their power, on this view, lies in their function: they are conceived of as ordering ideas, bringing unity into a mass of otherwise disconnected observations – as if what we see and hear were only amorphous stuff which stands in need of being fashioned by imposing on it certain forms of thought, such as substance and causality. Such at least was the idea envisaged by Kant and like-minded philosophers. Whether they were right or wrong in this, categories were certainly not meant to be classificatory labels. The theory of types, though not without points of contact with categories, is yet designed for altogether different ends, namely for the avoidance of antinomies in the logic of classes.

Categories in this new sense – of concepts necessary for unifying experience – seem to hold a peculiar attraction for philosophers: why should they otherwise be taken up so much with this subject? This may be explained in terms of a mistaken belief, not only in the 'correctness' but also in the universality of ordinary language; and this shows itself nowhere more clearly than in the trust in the applicability of the categories in the new sense. They were 'derived' (somewhat obscurely) from a table of judgment-forms; these, in their turn, were abstracted from ordinary word language – e.g. from the use of such words as 'all', 'some', 'if', 'or', 'is', 'may be', 'must be'. When seen in this light, the 'critical philosophy' reveals itself in the last analysis as the outcome of an even more deep-rooted mistake, namely as the result of a quite uncritical confidence in ordinary language.

Now language *is* a fashioner: not only does it carve out single facts from a sort of undifferentiated solid plentitude, but our whole awareness of them is tinged by the modes of expression in many subtle ways. So language does, up to a point, impress its forms on the way we perceive the world. Moreover, by making us pay heed to what is alike and what is unlike, what is a thing and what its properties, what is cause and effect, it interprets the world for us and even gives it a measure of comprehensibility. All this is due to the binding and congealing forms of language, and next to the existence of a vocabulary fit for describing, say, causal connexions. At this stage Kant made a fateful mistake: he absolutized the forms of language – or rather certain forms of it – by declaring them to have an *a priori* status, i.e. to be *necessary* conditions of any possible experience. These forms, the categories, came thus to be regarded as moulds of *the* – or should I rather say only of our specific human? – understanding, and therefore to be looked upon with a peculiar respect. (Oddly enough, Kant's table seems far from complete: key-notions such as 'equality' and 'similarity', 'continuity' and 'discreteness', 'finiteness' and 'infinity', 'probability' are missing, not to mention other blemishes.) The forms of language, whether in Kant's, or in a broader sense, were found to work successfully in describing, and partially in interpreting facts and happenings of the outside world. In consequence, the trust which these forms had won in the range to which they were applied was extended into regions where it could no longer be submitted to any practical check – indeed, to everything real. Thus a new philosophical attitude began to grow up in which space, time, and the categories were supposed, in some mysterious way, to be imprinted on anything that can ever become subject of experience; so that, on this view, whatever may enter our experience must already be patterned in a quite definite way. In other words, the natural facts themselves must conform to those laws the philosopher had found in the structure of thought or language.

Many of Kant's adherents thus fell into a superstitious attitude with regard to categories: these were said to exist independently of the real world and thought of as a sort of iron – no, superhard – grooves in which our thought is for ever and inescapably fixed as though by a spell. An instance of this reckless confidence in the power of categories is to say, 'Every event has a cause'. This sort of statement was claimed to embody an 'ideal' or an *a priori* truth which no empirical investigation can ever shake, based, as it was claimed, on the 'forms of understanding'. It was therefore said to hold universally; not only of the perceptual world but of all spheres of existence, for instance, of events of the atomic scale – a belief that died at about 1926. At the same time substance, too, went the way of all flesh.

In retrospect it becomes clear that two mistakes join hands here. The first, peculiar to Kant, is to project certain forms of language into reality; this has about the same effect as an optical trick: we are straining our eyes to the utmost to catch a glimpse of what is merely engraved upon our glasses. The second is the belief that the forms of natural language, evolved in practical life as they are, will be applicable outside their original domain. For this, there is no guarantee. On the contrary, with the rise of new modes of thought in the field of physics it dawns upon us that science transcends all forms of word language, just as it transcends ordinary thought.

Very different is another question which has caused philosophers a good deal of worry–'What is a rational decision?'[3] Was it, for instance, rational of Columbus when, after a few weeks sailing towards the west, he persisted in his course despite the danger of mutiny and the likelihood of food and water giving out? The problem is how to decide, in the absence of sufficiently complete information, as to the consequences of the possible alternative actions. Well, and how is one to decide? The difficulty seems to be how to weigh the pros and cons, infected with uncertainty as they are. Lately mathematics have begun to build up a theory of 'rational' decision-making in problems presenting such uncertainties. It has grown out of attempts to apply general principles of 'rational' conduct to games of pure chance, where recourse is had to probability theory and statistical inference. Can such a theory be brought to bear upon the present question? The answer depends on a number of assumptions. In the first place, one has to associate a certain amount of 'satisfaction' with the possible outcome of a gamble, an amount that may assume a positive or negative value. But are not satisfaction, discomfort, risk just the sort of thing that defy precise comparison, let alone measurement? The fact is that such measurement is possible if one is willing to accept certain postulates. On this assumption it can now be shown that, even if Columbus had no idea as to the chances of land being near, there would still have been several lines open to him: he might have followed the 'principle of insufficient reason', the policy of 'visualize the best', or that of 'visualize the worst' (also called 'minimax' because it amounts to minimizing the maximum risk). If certain definite figures are assigned, as 'satisfaction values', to the possible consequences of his decision (such as 'prospect of glory', 'prospect of death', 'prospect of glory and death'), it would have been rational for him to turn back in the first and third case, and keep going in the second. If, however, certain other figures are chosen (which might be considered more realistic), we get a different result: on this assumption it would have been worth-while for him to keep going, no matter which of the three principles is applied. That is to say, one and the same decision may appear 'rational' when judged from the one set of assumptions, and 'non-rational' when judged from the other. Thus a double uncertainty is involved in applying a theory of this sort – one springing from the arbitrariness of the figures assigned to the possible consequences (of his turning back or going on in case land is near or not near), the other from there being at least three different principles each of which has some claim to being 'rational'. The last remark may seem odd and contrary to expectation. Is there, after all, not *one* principle which embodies the idea of 'rationality'? This was for a time a moot point. Attempts have indeed been made to discover logical paradoxes such as would eliminate one or another of the principles from the competition; the remarkable thing is that they have all failed. There is at least one thing to be said for making mistakes and failing: if you do, it starts you thinking. And so it went with the attempts of discovering *the* 'rational' principle. On the contrary, it has been argued that even a principle of 'mixed' or 'randomized' strategy is reasonable, the idea being to let chance play a part in coming to a decision – reminiscent of the old and for many reasons venerable custom of counting one's buttons when faced with an awkward decision. Examples can indeed be produced to show that such a strategy – apparently the very opposite of a rational one – may yield better results than the best 'pure' strategy. One obvious advantage – important to generals, it would seem – is that it is spy-proof. To return to Columbus, however – whether his decision was 'rational' or not depends in the last analysis on the assumptions which underlie the whole calculation. As none of them can be proved to be the correct,

or even the most natural one, the question cannot be settled, not conclusively: here, then, at our first step, we are already faced with an *undecidable
issue*. The upshot – that, given a concrete situation, it may not be possible
to determine what is, and what is not rational – is not arrived at by analysing the use of language: it is the result of applying a mathematical theory
which, as such, is beyond the range of linguistic analysis. I do not mean to
say that language does at no stage of the argument come in – a certain appeal to it will be inevitable if only in discussing the rationality of the principles; but even so, it merely skirts the fringe of the subject, while leaving the
decision to considerations of a very different kind.

NOTES

[1] Ludwig Wittgenstein, *Philosophical Investigations, § 383.*
[2] ['spitted' (or 'stuck') 'on a pin' is perhaps meant.]
[3] [The placing of the section that follows is an editorial conjecture.]

CHAPTER XIII

BELIEF AND KNOWLEDGE*

As regards the triangle 'knowing – doubting – believing', it gives rise to the most variegated, not to say contradictory relations. One may believe a thing, merely believe, because one is not too sure of it, and one is not too sure of it because one has some doubts. If there were no doubt, one would not believe, one would know. Can anything be plainer than that? Yet it seems equally plain to say that one believes because one does not doubt, since if one did doubt one would not believe. Thus one may believe because one does not doubt, and one may believe because one does. But if anyone were to conclude from this that to have no doubts about a thing amounts, if not precisely to knowing all about it, yet at least to entertaining some sort of belief, he would just be shutting his eyes in the face of truth. For in fact it is quite easy to be without doubts or suspicions of any sort, to be guileless and yet utterly to believe – as in the case of the young man, who reading the telegram, says, 'There's no doubt I have passed – but I *cannot believe* it'.

One would perhaps expect knowing and believing to make common front against doubting. But nothing could be more wrong than this, and everyone can see that for himself when he remembers the simple truth that with knowing doubt only grows while with believing it melts away like snow in the March sun. Just how queer the situation is appears perhaps from the fact that a case can easily be made out both for saying, 'The less one knows the more ignorant one is' and for saying, 'The more one knows the more ignorant one is'. Indeed, to amplify only the latter, the former being truistic, imagine a man walking into a University Library, or the British Museum, and looking in awe at the treasures of information piled up along the ranks. The deeper he penetrates into this storehouse of knowledge, the more must the unlimited extent of his own ignorance dawn upon him. But to grasp it fully is denied him. For with every advance of knowledge there goes hand in hand an increase of uncertainty, as every expert will attest to,

* Composed in English, apparently in the 1950's.

and therefore of ignorance; so that it is safe to say that ignorance is a state – most finely suspended between the public and the experts – which advances day and night, indeed from hour to hour, while one is simply sitting there doing nothing. And what is even more remarkable, you cannot do anything about it: indeed, the more cruelly you try to persecute ignorance, to stamp it out root and branch – by pushing back the frontiers of the unknown in every possible direction – the better only will it flourish so that in the end it will rise triumphantly from all trials and tribulations.

1. BELIEVING

As certainty is related to knowing roughly as uncertainty is to doubting and believing, as least a few words must be said on these topics, if only as gleanings after the reapers have left the field.

To start with the word 'believe': its meaning glides on the one hand into 'thinking', 'supposing', on the other into 'having confidence' or 'faith' in a person, and consequently, into 'trusting', 'relying upon', and hence further into 'giving credence to' (a person, or his statement), but also into 'expecting', 'hoping', 'being afraid of', 'fancying' – senses grouping themselves more or less about the central meaning of 'to be of opinion', 'to hold it as true', 'to accept a statement as true', 'to acquiesce in its truth', etc. Even so, the word shifts over a wide range of meanings – from 'being convinced' to merely 'having a certain (vague) impression'. Historically, the nuclear sense is by no means the original one, being derived from 'to have confidence or faith in a person'. This, the oldest sense, still colours the wider use of the verb.

To mention a somewhat similar case, the word 'true' (cogn. with 'troth', 'betrothed', G. *treu*) has for its original meaning, i.e. so far as written records go, (a) of things: 'reliable', 'constant', 'sure', 'secure'; (b) of persons: 'steadfast in adherence' to a commander or friend; to a cause or principle; to one's faith, promises, etc.; 'faithful', 'loyal', 'trusty'; (c) in a more general sense: 'honest', 'upright', 'trustworthy', 'free from deceit', 'sincere', 'truthful'. It was not before fully four centuries later that it was transfered to statements and acquired its present sense. Much of its overtones, especially the strong moral one[1], is the result of what it has been through. In particular, what has been called its 'performatory' role, not being rooted in the (present) meaning of the word, is peculiar to English, not to the concept of truth as such, though traces of it are to be found in other languages too – a warning that keeping to one's mother tongue may not be without danger in philosophy.

As with 'true', so with 'believe'. Behind the present meaning of the verb is ranged another and graver one, more or less distinctly felt. Thus it comes that, for all the variegated later use, the original meaning never glimmers out of sight altogether, being merely relegated to the background, and always ready to come to the fore – for instance, and above all, in religious language. This fact, that the primary sense accompanies any of the present ones like the sympathetic chimes of a bell, or acts as an echo to them, explains what any sensitive observer will feel: something of a multiplicity of meaning, an indefiniteness which we shall do well to bear in mind.

But apart from the dictionary distinctions referred to there are other ones. Even when taken in its most central sense, 'I believe' can mean many different things. To begin with, there are two aspects, active and passive. 'I believe that the earth is round' is an example of the first kind; it means, as commonly understood, that I have learned this sort of thing at school, perhaps without paying much heed to the reasons given or examining them closely. In all that I was predominantly passive, like a moist sponge filled with the little facts it has soaked up. 'It is my belief that sentimentality and cruelty are both on the increase today' is in a different case: I feel convinced of it, say, as a consequence of certain observations and reflections of my own, I can state my reasons and not merely ooze them out, in short, belief here is the result of an activity: it is *my* belief.

But this is not all to be said about it. Just as it may happen that I do not speak but only hear myself utter certain words, so it may be that I do not believe but find that a belief has formed itself without waiting for the originator; as for me. I may be left puzzled as to the reasons for it. Such a belief is 'mine' and yet not 'mine' – not so unlike the inspiration ascribed to a poet. And just as a poem is made by two, the author and what is in him, so a belief of this sort is neither personal nor yet non-personal, rather belonging to that twilight zone of the self where the whole antithesis becomes blurred. Beliefs, like wishes and intentions, are not always, or not fully, under conscious control – a hint that there may be some closer relationship between them.

There are other differences beside that. If I say, 'I believe', am I saying something about myself? I may, or I may not. I am letting out something about myself if I say, 'I believe in ghosts', and this would have come out even more forcibly had I said, 'I am quite willing to believe in ghosts'; nothing of the sort is involved if I say, guardedly, 'That, I believe, is a myrmeleonina'. Not that these two senses are always clearly separable. Considering a dictum of Einstein's, "The most incomprehensible thing about the world is that it is comprehensible", what is it I believe in when I say, concurring, 'Yes, I do believe that the world is comprehensible'? Can I state it more, or quite precisely? When I try to do so I may well, and for all I know, most probably fail. And if so, I am liable to be taken to task. ('Look here, you did not really mean anything at all', as if I had been talking froth.) Part of what is meant, I suppose, is that nature obeys laws of a simple or harmonious kind, or laws that are rational, transparent to reason. (But what is 'simple', 'harmonious', 'rational'? It is far easier to get the 'feel' of it in practice than to condense it into a workable definition.) What seems more important, however, is that 'believe', as used here, is indicative of a certain confident frame of mind. After all, just as one cannot jump or win a race unless one has confidence, so one cannot be a discoverer without the anticipatory belief that the goal is within reach. So construed, belief is a power that may give a man like Kepler or Einstein the strength to persevere in defiance of failures and reverses, and this in spite of its being very uncertain whether its content can ever be formulated adequately. My statement is then in part an assertion about nature, though an exceedingly hazy one in the light of which one sees about as clearly as at night with the moon behind the clouds, and in part a disclosure of one's own attitude of mind.

Another difference: a belief may be fully or only half-verbalized, or else wholly unverbalized. An idea of what belief is in the last case may be gained from considering the behaviour of a man being forcibly dragged into a fire[2]. He will resist with all his might, strike out with hands and feet like a madman. He is deadly afraid, and this state, being afraid, is certainly much akin to belief. Now does he act so for any *reasons*? Does he, for example, remember bad experiences in the past–think e.g. of induction, say to himself, 'Ah, of course, the principle of concomitant variation', or something of the sort, and then conclude that, after all, the probability is very high that the fire will burn him? Does he? He strikes out like a madman; well, this *is* belief, inarticulate belief.[2] (And this sort, previous to language, may well exist in animals.)

'To believe' often means no more than to be in a certain trusting, hopeful, or expectant frame of mind, or else to be uneasy or alarmed, postures which for the most part remain *half*-formulated. In other cases, however, believing is brought in contrast with such feeling-tones, as in Lichtenberg's remark, "He not only did not believe in ghosts, he was not even afraid of them."

One may believe something without being able to say why, indeed in spite of evidence to the contrary, brushing it off with that contempt that has earned it the epithet 'sovereign' – as when a gambler says, 'This is my lucky day', or when a man is imbued with the conviction to be chosen for some great purpose, though he could not for the world say what it is; or again when one has that sort of foreboding which makes people say, 'I feel it in my bones' (that something untoward is going to happen). Women often have this sort of belief; they are then fond of saying that they are being 'irrational' – as if they were able to. (Another favourite with them, of course, is 'intuition'). Everyone can see the marked difference between this type of believing and the more common one; for one thing, it is immune to reasoning ('He doggedly believes that: you cannot talk him out of it').

As an example of an 'irrational' belief – if there is such a thing – consider the last minutes of a prisoner sentenced to death as described by Dostoevsky: "He gazes at the guilded cupolas of a near-by church, at the sunbeams reflected by them, and imagines that they represent the new form of being on which he is about to enter. The belief that in three minutes his being will be fused with theirs moves him to horrified repugnance." Mention should, perhaps, also be made of the occurence of certain queer belief-states such as described by James. According to him, a person may, in a state of intoxication, have a feeling of conviction heightened to an abnormal degree, and yet be totally unable to say what he is convinced of. If this is correct, it would seem that a conviction – or how shall I put it, a feeling of being convinced, a propensity to believe? – may exist unattached to any content, free floating as it were, and apt to pounce upon any – suitable? – content.

But to return to less strange things, there are certain states of mind in which one believes and yet does not believe, half-believes. "At that moment I believed myself transported back to the past, as though by magic to re-live an hour of it and breathe in once again the fragrance of my aunt's tea". Did I believe it, or did I not? The word 'believe' is perhaps too pretentious an expression for something that is a mere breath. That the line is by no means always easy to draw, or may even be lost in uncertainty, is notorious. Thus

Freud once said of certain assumptions of his that he himself could not tell "how far he believed in them"; quite a common case in science.

Now for another point. According to a widely held view, 'to know' and 'to believe' are dispositional verbs, though of disparate types. To ask, 'At what time did you begin to believe?', or, 'Exactly at what moment did you finish knowing?' sounds wildly out of place (though I may, of course, say, ironically, 'Now I've finished knowing him'). From this it is concluded that these are 'non-episodic' verbs, i.e. expressions not answering to any specific 'acts', incidents or episodes going on in us while we e.g. believe something. Though this has gone far to do away with occult 'acts' and ditto 'states of mind', it is an over-simplified story. For the actual use of these verbs is more complex than that and does not quite fit the cake-moulds of the account. Take 'I knew it all the time'; a quite current expression like the amiable 'I always thought of you', and like the latter to be taken with a pinch of salt. For it would not be literally true to assert of the speaker that he kept thinking of you 'all the time', every minute, day and night, but rather, supposing he is sincere in what he is saying, that his thought 'tended', or was 'inclined' to turn to you, using associations as stepping-stones, and suchlike –in short, that it seized just about any honourable occasion of popping up. And much the same goes for the sentence, 'I knew it all the time'. Somewhat paraphrased, it may be read, 'I was all the time dimly aware of it (say) whenever the subject was touched, but I did not quite dare to tell it to myself in so many words'. An untold tale of suspense hangs on that, a suggestion of anxiety and suppression, depending on the context, so that there is quite a lot of the episodic to it. It is a big mistake to suppose that, because a verb does not refer to a *particular* occurrence, it has no reference to occurrences at all. The two expressions *are* no doubt dispositional, and yet not in the standard sense of 'whenever I would be asked I would say', i.e. not in the way in which the sentences 'Rubber is elastic' and 'I know German' are: they are partly dispositional, partly episodic, with more of the emphasis placed on the latter. 'At this moment I began to believe him': there is nothing wrong with that. Trust may come suddenly, just as its opposite, suspicion, may awake in an instant. It may be no more than a smile flickering in his eyes, a moment's hesitation, an undertone in his voice that makes – and marks – all the difference. 'At that instant I knew (believed) that I was in danger': i.e. at that instant the idea 'danger!' flashed through my mind – a use which shows the dispositional in full retreat and the episodic dominating the scene. What such examples are meant to illustrate is that there is an almost continuous line running from the one pole, the purely dispositional, to the opposite one, the purely episodic.

This makes us see another ambiguity. We speak as if we always believed that the earth is round, whereas we should distinguish between belief as a disposition – that we are ready to believe it when the occasion arises – and what may be called the 'real' or 'live' belief, when we are actually persuaded of the fact, 'feel sure' of it.

Another question: Does belief always lead to action? So far we have paid heed to the content of a belief only, i.e. to something which may be either true or false. But one may look upon a belief from a very different point of view, seeing in it something that acts like a force in that it is apt to discharge itself into action. That this is true of many types of belief – e.g. of political ones – will hardly be disputed: but is it a characteristic of any belief? Belief in non-resistance seems an obvious counter-example. Though it does not issue in 'action' in a very strict sense of the word, yet it is none the less true that such a belief will mould the whole conduct of a person or a people. A purely intellectual belief is a better case. If I believe that Goldbach's theorem is true, no action whatever will issue from that – except in the case that, being a mathematician myself, I may be incited to go in research for a proof of that conjecture. Now a 'dynamic' belief is obviously in some respects very unlike a non-dynamic one, and this brings out a further distinction.

Thus we are faced with a variety of senses of the verb 'believe'. It is almost as if different concepts were lodged in the same word-husk, or as if a number of small but fairly well-defined concepts were hanging loosely inside the hull of the big ill-defined one. Or is it presuming too much to speak here of 'concepts'? Ought I rather to speak of 'sub-meanings' only? I am not too sure. After all, what has to be added to a meaning to raise it to the dignity of a concept? The existence of a separate word for it? If so, this would make the whole distinction a contingent one, being dependent on the wealth of words in a given language – which may not be to everyone's taste. Yet a pretty clear token of that being so is to be found in those cases where some other European language offers us a choice of several words to express senses undiscriminated in English. 'To know' is a case in point: it occupies the ground formerly covered by different verbs (WIT and CAN or KEN), and it still answers to *connaître, savoir* in French, and *wissen, kennen,* and, in part, *erkennen, können* in German. In such a case at least it seems arguable that one English word holds in itself several distinct concepts.

Be this as it may, we have not yet reached the end of the variations in the uses of 'believe'. We are reminded of another one when we compare the two sentences 'I believe you' and 'I believe in you'. The latter one suggests that I believe, not only what you are saying this time, or indeed at any time, but above all that I have belief (trust) *in you* – whether in your integrity, goodness, career, destiny, or God-knows-what, this being left open (unless the situation provides a key); consequently it appears to some degree indefinite, being poised between several potential senses. Now where is this construction in place? The way we speak in ordinary life may be illustrated in this way. While we say, 'I believe his story', a child 'believes in fairy-tales', the believers 'in God' or 'in the Bible', and all of us 'in Shakespeare's characters', including Caliban. That there is a difference between these cases is quite obvious, though it may be less obvious how to express it. To throw out a hint, when a child is said to believe 'in' fairy-tales, part of what is meant is that he is not yet developed far enough to take them for what they are. That he is not (or hardly ever) said to believe in a *particular* fairy-tale suggests that it is rather the whole species that matters here. The point of this construction can now be seen: it serves to characterize the child's mental stage by saying something about his attitude towards such tales. As with the child, so with a believer. No one ever says that he 'believes the Bible', though it is, from a grammatical point of view, just as correct to say that he 'believes what is written in the Bible', as '*in* what is written in the Bible'; and it is correct because, in the first instance, the reference is to the content only, in the second, to Holy Writ itself with its aura of awe. 'To believe in the Bible', besides characterizing anyone's frame of mind, is indicative of a further point: the authority of the Bible and, what goes with it, its importance. Only where there is authority or importance is this construction appropriate. It would be very odd if someone were to speak of his belief 'in' such-and-such a book, although he may, of course, believe 'in' all sorts of doctrines, theories, principles expounded there; or else 'in' reading in general. Similarly we speak of a belief 'in the triumph of good over evil', 'in reason', etc. It seems fair to assume that we believe in anything we believe worth believing in. Modern man believes in individualism and in collectivism, in a strong navy and in pacifism, in nationalism and in world government, in old romantic ruins, moonlight and moors, and in stream-

lined factories, functionalism and symbolic logic, in science and in E.S.P.:
in short, in any idea and its counter-idea, as though mankind as a whole
were seized by a fit of schizophrenia. It is here that the primary sense of the
verb, the non-intellectual one, shines through clearly.

There is, however, another sense in which this construction is often used,
a sense which is not suggestive of authority or importance – as when we
speak of a child believing 'in fairy-tales', or of our belief 'in Shakespeare's
characters'. The last one is a very special sense of 'believing in', and one in
no way easy to analyse. What we mean by this, by a character 'we can believe in', seems, in some respects, to be different from 'true to life': it is a
character, we feel, that must be captivating to our imagination or at least
arouse our emotions, even if, like Caliban, he is not drawn from nature at
all, 'a species of himself': yet as 'living', 'convincing' as any of the human
characters in the play.

Next, it should be noticed that exactly the same words, when used paren-
thetically, perform a different job. 'I believe that the Bible is true' is – nearly
– the profession of a creed, while 'This man, I believe, is a crank' is a specimen of talking guardedly. The mere position of the words alters their sense.
And not *only* the position. Had I merely said, 'The Bible is believed to be
true', not the ghost of a profession would attach to it. Whether 'believe'
does, or does not, profess anything seems, then, besides position, to hinge
on personal construction. That even this is not quite to the point appears
from comparing the cases (1) 'Most things in the Bible, I believe, are true'
(non-committal) and (2) 'Since most of the Bible, as I believe, is true', (say,
Divine Revelation speaks through it): here the words inserted do express
a sort of faith.

To pursue such differences further is, however, not the object of this article. Only one more distinction may be noted. Supposing that a man believes 'in' Goldbach's theorem and another 'in' mathematics, is this belief in
the same sense? Here another difference comes to the surface, for while the
former refers no doubt to the truth of the conjecture, the latter is far more
indefinite: it may mean belief 'in the truth', but much more likely 'in the significance' of mathematics, with lots of questions swarming round that.

Reference has already been made to a concept closely akin to 'belief',
namely, 'faith'. In speaking of fiath, I shall confine myself to its non-the-
ological aspects. Now faith – in the sense of 'trust', 'confidence' – is, generally speaking, less intellectual than belief: it is more often the expression of

a person's attitude, whether with regard to another person, a cause, a doctrine, or a way of looking at things. It is the latter of these senses which will concern us here. Now what, in this case, is an 'attitude'? We are, most of us, under the influence of certain ideas which, so to speak, are 'in the air'.

[The remainder of this discussion, if it was ever composed, is now lost.]

II. KNOWING

This brings me to the next subject, 'knowing'. Is knowing merely a particularly high degree of believing? Or is it different from it in principle? If so, in what does the difference consist? Before descending into the arena, however, a general remark. The first thing that strikes us is the fact that in talking loosely, there is not much difference between the two terms. Thus I might say, 'I know he is in' where, more precisely, I should have said, 'I believe he is in', 'To all appearance he is in', or something of the sort. Now I cannot bring myself to regard this loose way of talking as such a terrible thing, considering that we can never express ourselves *quite* precisely. Moreover, there are in this case all sorts of other considerations which may have a claim to our attention, such as questions of politeness, etiquette, and prestige, and, above all, tact. Churchill, in a letter to Roosevelt: "I venture to put this point before you although I know you must understand it as well as I do". 'Believe' would here be quite out of place. Another point: persons of rank are not supposed to believe: they 'know'. Even a quite common G.P. would say, in that confident calm which only superior knowledge can confer, 'Take this stuff, my man, 3 times a day, I'm sure it will do you a lot of good', whereas all a research scientist might say is perhaps 'Humph!' For all I know, politicians as a whole (I am not talking of individuals) seem to have no particular liking for the verb 'believe', e.g. when addressing mass meetings: they believe, perhaps, it may not be too good in election times, and they know how to avoid it. And I would not blame them. After all, there is really something weakish about the word 'believe', a certain uncertainty, a toning-down quality which the speaker may have reason to eschew. Nor can it be said that all of us are all the time above such worldly promptings; not to mention the fact that simply to disregard these promptings would mean to shut one's eyes to one of the forces which shape living speech. For these and other reasons it is not the *looseness* which is to be blamed, not primarily; it rather points to other forces behind the scene which subtly influence the usage, bending it this way or that.

If, as I have been suggesting, there is something to be said for the 'loose way' of talking, I do not wish to give the impression that Lam pleading for licence. I do not: I do not deny that there is a difference between knowing and believing, and, in a wide range of cases, one in *kind* and not merely in degree. But the discrimination is one of those nice points which require for their settlement a perceptive mind rather than precise or ready-made rules. It is all a matter of a delicate sense of what is fitting and what is not. But tact and sensibility are not entirely arbitrary matters; they seem to be guided by a sort of insight; only I find it highly questionable whether this insight, and the linguistic practice which issues from it, can ever be distilled into clear rules.

But let me turn to more concrete matters. It is where knowing and believing make contact that a number of new questions arise. Knowing is as a rule contrasted with believing. But this is altogether too simple an account. The point is that the verbs 'believe' and 'know' are embedded in language in a different way, with different constructions and ways of entering into combinations with other words. A long list of phrases has to be written out and compared before one can trace the curiously jagged borderline. Here are some scanty examples.

One can know, not believe 'accurately', 'for certain', 'officially', whereas one may believe but not know 'strongly'. Roughly, belief is related to inclination words so that, while I cannot say, 'I incline, or tend, to know', I may say, 'I incline, or tend, to believe'. A man may come to know himself, not to believe himself, though perhaps to believe *in* himself. Though one may ask, 'How do you know?' but not 'How do you believe?', one may say, reproachfully, 'How *can* (*could*) you believe such a thing?', or 'But how *can* you be sure?' In some cases, replacing 'knowing' with 'believing' alters the sense altogether: 'I know him well', 'I well believe him'; 'What the truth is we shall never know', 'What the truth is we shall never believe'. In other cases, it turns sense into nonsense: 'I know French', 'I am in the position to know', 'I believe so, but I may be wrong'.

In the main, one has rather the impression that our whole attitude is different in both cases – which comes out, perhaps, in the difference of construction: I know *of*, or *about* something, but believe *in* something – as if my relation to the object itself was not quite the same in the two cases. And this is not just an idiosyncrasy of English, for this feature is paralleled in other language: *von, um etwas wissen, an etwas glauben.* Of course, there are a good many other constructions in which the two verbs occur in perfectly symmetrical ways. Thus the appeal to the use of language, it would seem, is inconclusive.

Next, there are asymmetries of a peculiar sort. Thus it makes sense to say, 'I know that so-and-so is the case' and also 'I *know* that I know that –', though,. if one were to go on in this series, after a few steps the sense of the string of words, 'I know that I know that –' becomes increasingly nebulous; similarly, beginning with, 'I believe that –', we may proceed to 'I *know* that I believe that –'. But there is no use for an expression like 'I *believe* that I believe'. Yet, so odd is language, even that is not altogether true. "If men did not believe this in the strict sense of the word, they still believed that they believed it." (H. Spencer) Which has the effect of somewhat toning down the assertion. And as for the other phrase, 'I believe that I know', I may say, and quite understandably, 'I believe that I know how to do this sum', 'how to eat asparagus'. It should be noticed, however, that such difficulties, where they occur, are peculiar to the case that both verbs refer to the same subject: there is nothing obscure about saying, 'I believe that *he* believes', just as little as there is about 'I know (I believe) that *he* knows'. In certain talks concerning religious matters one might even say, 'I think that I believe', betraying hesitance, or else the beginning of conversion. (Tone of voice and facial expression must do the rest.)

Nor is this the end of the story. For when you say, 'I believe that I know so-and-so', you may be asked, 'Are you sure?' After all, believing has business relations with wishing, just as intending has with hoping – as brought out in the phrase 'wishful thinking'. ("Some negroes who believe the resurrection, think that they shall rise white". Sir Thomas Browne.) Now can you always discriminate between believing and wishing? But in casting such doubts on the matter I might be going just a bit too far. For though there may be occasions where such qualms arise naturally enough, for the most part they do not. And to carry suspicion to its extreme here is as if one doubted the difference between red and yellow because of the existence of orange. There is a danger of being too clever. Moreover, to ask a man, 'Are you *quite sure* that you believe?' only lures him into the wrong sort of response. For it makes it look as if he did not quite know and had now to make sure by applying some sort of test – say, by peering hard inside to discover there whether he really knows. This only encourages a wrong sort of introspection. To ask a man, 'How do you know that you believe (have got a toothache)?' etc., or 'Are you sure that you believe?' etc. is a misleading question as well as a leading one. (A case for linguists, it would seem.)

To turn now to some more difficult cases: can anyone believe a thing and, at the same time, believe the opposite? I think this is possible.

[The remainder of this discussion also is lacking.]

NOTES

[1] To say 'That is not true' ('untrue') of something told or reported is almost a personal insult in English; not so in German, as *wahr*, being influenced by Lat. *verus*, has not shared the career of 'true'.

[2] [The example (which also occurs in chapter V above) is drawn from a dictation of Wittgenstein's of *ca.* 1931: '*Diktat für Schlick*'.]

CHAPTER XIV

TWO ACCOUNTS OF KNOWING*

Of particular importance – because something in philosophy hangs on it – is the question. Is knowing a limiting case of believing? Here I want to take a definite stand by saying: Yes, it may – or it may not.[1] Let me contrast two cases.

I. KNOWING AS A LIMITING CASE OF BELIEVING

Suppose, as has happened in war, that it filters through to the men in a combat unit that they are trapped and that it is hopeless to fight on. There is a certain situation in which they *know*. There were at first perhaps only rumours, then one man or the other saw that they were encircled, and in the end they all know. Here suspicion, rumour, belief changes, by degrees or abruptly, into certainty. In such a case I would say that knowing is a limiting case of the other weaker forms. Whether the first or third person is used in recording the fact is of no consequence.

Let me now take another case, culled from science, which points the same way. I choose an example from science because it shows the whole situation against a somewhat different background.

Not so long ago, in the 1940's, hundreds of newborn babies were hit by a mysterious kind of disease – they were going blind for some unknown cause. One of the things that was puzzling about it was that the babies afflicted had all been born prematurely; another that the epidemic tended to spare some hospitals, some towns and even whole countries – though it turned up in widely different places, in England, in U.S., France and Australia. Some doctors began to suspect *oxygen deficiency* to be at the root of the trouble, a view suggested by certain features of the attack: it was a 'systemic disease', i.e. a general physiological disorder such as had been observed often to go with lack of oxygen, one which attacked not only the eyes but other organs and tissues as well. Accordingly infants were treated with

* Composed in English, apparently in the 1950's.

this life-giving substance, but oddly enough they got only worse. Suspicion fell now on all sorts of things–blood transfusions, anaesthetics, vitamins, sedatives. Then one day someone began to wonder whether an *excess* of oxygen might not be responsible for this kind of blindness among babies, though why it should be so was unclear. In the meantime it was found that oxygen may indeed have a toxic effect no less than a therapeutic one. Airmen and mountain climbers, for instance, had noticed that breathing pure oxygen in lower altitudes produced burns and other things. Thus little by little there emerged the paradoxical truth that it was precisely by giving *too much* oxygen that a shortage of oxygen in the blood was caused: the infants were starving in the midst of plenty. It goes without saying that much more clinical observations, experiments on animals etc. were needed, before all the pieces of the jig-saw puzzle were fitted together and the whole pattern could be seen.

Now about the time when the first clue was found (which, by the way, was no more than a statistical observation) a doctor might have said, 'I should not be surprised if an overdosage of oxygen was the trouble' – to add after evidence had been accumulating, 'Now I *know* (that this is so)'. As the strange story of the discovery unfolds, and one piece of evidence after the other is seen to emerge, revealing more and more of a very unusual causal nexus, it becomes idle to speculate at which point the idea was merely a suspicion and at which it turned into certainty. Plainly there is a continuous scale, running from the merest 'I have got a hunch' or 'I suspect' to 'I am pretty, quite sure' and 'I know'. Even the last case is not *sharply* separated from the other ones, especially as knowing, in the sense it is used in science, carries no suggestion of infallibility. Knowing *is* a limiting case here. However, as science is team-work, the common form is not so much to say '*I* know' as 'It is known' – in itself a warning to[2] interpret the verb without ado in a performatory sense. And even that may at times go further than is warranted. No scientist is likely to say, 'It is known how life started on earth' – just like this; what he would say is perhaps, 'There are good reasons for supposing –' or something of the sort. At this point the sad fact can no longer be hushed up that scientists as a whole show a marked repugnance to use not only the term 'know' but also 'true', 'prove' and 'certain'. To cite just one example that may stand for many: in that masterly and thorough account of relativity written by Pauli – a highly critical study, comparing Einstein's theory with a number of rival ones where every word had to be weighed like gold – in that treatise the word 'true' does not occur at all, or, to be precise, it does occur in one single passage, and that is the sentence, "The question which geometry is the true one is devoid of meaning". Now what exactly have scientists got against this semantic group? Why do they reject the idea of using any of those terms as if it were a personal insult? Not from bashfullness, I should think, nor from an undue desire to express themselves as guardedly as possible; no, it is not quite that. The fact is that the members of this group are, all of them, just a bit *too* definite and, if I may say so, too crude for the scientist's purposes. Instead, a quite elaborate idiom has been evolved with very fine gradations for bringing out all sorts of subtle shades. By the way, it must not be assumed that the shades on this palette form a linear series: it is much more complex as is, after all, only to be expected, considering that a scientific theory, indeed truth itself is such a many-facted thing. This suggests a subject too vast to be surveyed in a single paper – the way scientists talk about their theories and hypotheses and appraise their strength or weakness.

So much about knowing as a limiting case of believing. I must turn now to the second view, namely,

II. KNOWING AS BEING ENTIRELY DIFFERENT
FROM BELIEVING

Here we enter upon Professor Austin's province. I am a great admirer of his paper on this subject[3] which in my view, constitutes a major discovery. But on thinking the matter over I have come to find myself in disagreement with him on a number of points, and I think that I should mention one or two of them. According to Professor Austin, there is a big difference between the two phrases 'to know' and 'to be quite sure', even 'to be absolutely sure', the ground being that in the former case I take *responsibility* upon me ('give you my word' for it). So far as I can see, this is only *part* of the story.

In the first place, it should be noticed that the utterances, 'I know', 'I promise' do not always carry a performative load. When I say, with a twinkle in my eyes, 'I promise you fine weather for tomorrow's party', or, smilingly, 'I know what you are up to', only a fool would take me at my word. It does not hinge on the *words* but on the way they are embedded in a situation. Irony, for instance, has a way of twisting a sense into its very opposite, and it is not imperatively necessary to go to Egyptian to find that in most of our words there is dormant an antithetical double sense.

But even disregarding such cases and sticking to quite plain ordinary English, there are plenty of instances where the utterance: 'I know' is far from having any performatory over- or undertones. Dialogue: 'I was only joking'. – 'I know you were, my dear'. Or this case: 'I know you are my friend, I know that I can trust you': performatory?

Next, even the innocent and spotless looking word 'believe' may have something up its sleeve. Take, for instance, the case of a man who openly avows his belief in communism: he, too, is 'doing' something with words – abiding the odium that clings to this view, inciting others, etc. – and his utterance may even entail moral and legal consequences, just as in the case of the early Christians. Or, to change the example, suppose that we were living in a Catholic country, and that a man, in order to get appointed to some job, would have to go through a certain ceremony, involving the utterance of the formula 'I believe in the teachings of the Holy Catholic Church'. Now when asked by His Worship, 'Do you believe – ?' and he says, 'Yes, I do', while in fact he does not, he is misleading the good people, indeed playing a vile trick on them, yet not by using the form 'I know' but 'I believe'. That he is performing something with words is plain, and if Austin means nothing more than that I fully agree with him. Why, then, his insistence on the form 'I know'?

I want now to discuss another case where, if I see rightly, the utterance of the words 'I know' does not mean, or imply, 'I give you my word for it', where, in fact, this interpretation would be quite beside the point. The example I have in mind is of interest for yet another reason: it calls attention to a class of statements which are absolutely certain.

Suppose I say, in reply to a question, 'I know where I spent the summer of 1936'. Suppose, for the sake of argument, that this were called in question, or that it were suggested to me that my memory may play me false, then I should find myself in one of two situations: (a) I am really a tiny bit uncertain or become so under cross-examination; (b) I really know. It is the second case I wish to consider here. Well, if I *know*. No objection would make the least impression upon me; and even if evidence were produced against me (such as passport stamps, snapshots taken at the time and showing me on the *Canal Grande*, or heaven knows what), I would say, 'Faked!' No power on earth could shake my certainty – I had sooner think that the whole world must have gone mad, or that there is some conspiracy against me than yield one inch from what I know to be true. I would not dream of saying, 'I believe that I spent the summer –' or 'I am sure of it', or even 'quite sure', insofar as this would be admitting a ghost of uncertainty. No, I know, know, *know!* This is the only possible word for me to use here – I have no choice.

Now this brings me to the first point I want to make. Professor Austin, in his symposium article, says, "Whenever I say I know, I am always liable to be taken to claim that, in a certain sense appropriate to the kind of statement, ... I am able to *prove* it". Now I am not quite sure what he means by this – whether that I *may* be taken to claim, even though this interpretation may (perhaps) be the wrong one, or whether I *should* (in any case) be taken to claim, this being the right interpretation. However that may be, the point of my example is this that in this case I *know* and yet *cannot prove* it as all the people who might have testified to it are by now dead. Witnesses, documentary or circumstantial evidence – all right, but quite beside the point so far as *I* am concerned. I know what I know, *not* because it is derived from any such sources nor, of course, because it is supported by any evidence. Indeed, external conditions may be such that no recourse to 'evidence' is possible: thus I fully remember the day, it was my third birthday, when the idea of time dawned upon me, or rather when, for the first time, I became fully aware of what an enormous length one year is, being quite taken aback by this discovery and feeling old as the hills, or the moment I discovered the beauty of Mozart – inner experiences which, given the circumstances as they were, preclude any sort of corroboration.

But to return to the simple question where I spent the summer of 1936, this is quite a normal case and yet remarkable in this that in saying 'I know' I am not giving you my word: I just express what I know and impart to you a piece of information; everything else is beside the point.

Of course, it is different, when I do not *really* know, however 'sure' I may be. In such a case, if I am honest, I would not mind listening to doubts and objections, engaging in arguments, considering or sifting evidence, and all sorts of similar things. In short, I would be open-minded, ready to withdraw or modify my claim under the pressure of counterevidence, and all that.

In what, then, lies here the difference between knowing and believing? Not in the way I am using *words*, nor in the circumstance that different (moral or legal) consequences may ensue. No, the difference lies in the *whole way I behave.* In the one case I am stubborn, impervious to argument, unwilling to listen, obdurate; in the other I am not. This, I submit, is the real point that marks the difference between knowing and believing, or being ever so sure, and the rest – a point, however, that refuses to come out either in language or in law. In behaving as obstinately as I do in the example imagined – am I unreasonable, I wonder? Not a bit: I have a perfectly good right to brush aside such aspersions for what they are.

So long as one is talking in general terms, everyone, of course, agrees that anyone might 'just be mistaken'. But turn to a concrete situation, and everything is changed. If a man is married, it *just makes no sense* to talk of the possibility of his being mistaken. Possibility indeed! Eye-wash, that is what it is. And so in this case before us: all the philosophers' talk cannot in the least shake the fact that I know this as definitely as anything *can* be known. Every one of you knows hundreds of facts about which there is not the least doubt – facts such as that he is married, what his name is, etc. What, a mistake? *What is certain* if these things are not? They are, in a very real sense, the *prototype* of all certainty.

We can see now what is being confused here. The philosopher says, 'Yes, a statement *of this sort*, a statement *in every respect like this one* is (generally speaking) exposed to doubt'. The obvious reply is: But we are not talking of a statement 'of this sort', but of *this* quite particular concrete statement, and of *these* quite particular concrete circumstances. What right have you got to transfer what holds, or may hold, of statements 'of this sort' to *this* statement? None whatsoever! It is a complete travesty of the truth to cast doubts on this particular statement. What does the argument amount to? Simply to this: *some* statements of the sort have been found to be mistaken, therefore *all* statements of this sort are suspect. It is as if one argued: Women, in general, are... *La donna è mobile:* so this particular woman is .... It is outrageous, against all reason and logic! Philosophers have got so entangled in their niceties that they can no longer see the wood for the trees.

In this sense, then, there are particular concrete statements which are true beyond doubt, 'incorrigible'. (But they are very different from those mostly put forward as candidates, e.g. 'Here is something red'. To be married has, under the proper circumstances, much more claim to being indubitable than 'sense-datum statements'.) And this, I submit, is an aspect of 'knowing v. believing' that is no less worthy of our attention than the difference in performatory function – if there is any.

NOTES

[1] [This seems to be a fuller and later discussion of some points raised in the preceding paper.]

[2] [Perhaps 'not to interpret' is meant.]

[3] [The reference is to J.L. Austin's contribution to the symposium on'Other Minds', *Aristotelian Society Supplementary Volume* **20** (1946), frequently reprinted, as for example in Austin's *Philosophical Papers* 1961, p. 44.]

BIBLIOGRAPHY OF WORKS BY
FRIEDRICH WAISMANN

[1] 'Die Natur des Reduzibilitätsaxioms', *Monatshefte für Mathematik und Physik* **35**, 143-146, 1928. Reprinted in: [31], 1-3. English translation: 'The Nature of the Axiom of Reducibility' (in this volume).

[2] 'Thesen' (ca. 1930), privately circulated, first printed in [28], 233-261.

[3] 'Logische Analyse des Wahrscheinlichkeitsbegriffs,' *Erkenntnis* **1**, 228-248, Leipzig. Felix Meiner, 1930/31. Reprinted in: [31], 4-24. English translation: 'A Logical Analysis of the Concept of Probability' (in this volume).

[4] 'Über den Begriff der Identität', *Erkenntnis* **6**, 56-64, Leipzig, Felix Meiner, 1936. Reprinted in: [31], 25-33. English translation: 'The Concept of Identity' (in this volume).

[5] *Einführung in das mathematische Denken: Die Begriffsbildung der modernen Mathematik* (preface by Karl Menger), Wien, Springer, 1936; 2nd edition, Wien, 1947; 3rd edition (edited by Friedrich Kur), München, dtv, 1970. Italian translation: *Introduzione al pensiero matematico: La formazione dei concetti nella matematica moderna* (transl. by L. Geymonat), Torino, 1939; 5th edition with a preface by Corrado Mangione, Torino, 1971. English translation: *Introduction to Mathematical Thinking: The Formation of Concepts in Modern Mathematics* (transl. by Th. J. Benac, Foreword by Karl Menger), New York, 1951.

[6] 'De Betekenis van Moritz Schlick voor de Wijsbegeerte', *Synthese* **1**, 361-370, 1936. English translation: 'Moritz Schlick's Significance for Philosophy' (in this volume).

[7] 'Über Hypothesen' (before 1936?), [32], 612-642. English translation: 'Hypotheses' (in this volume).

[8] 'Ist die Logik eine deduktive Theorie?', *Erkenntnis* **7**, 274-281, 375, 1938. Reprinted in: [31], 34-41. English translation: 'Is Logic a Deductive Theory?' (in this volume).

[9] 'The Relevance of Psychology to Logic,' *Proceedings of the Aristotelian Society*, Supp. Vol. XVII, 54-68, 1938. Reprinted in: *Readings in Philosophical Analysis* by H. Feigl and W. Sellars), 211-221, New York, Appleton - Century - Crofts, 1949; and in this volume. German translation: 'Die Relevanz der Psychologie für die Logik', [31], 67-80.

[10] Vorwort, Moritz Schlick, *Gesammelte Aufsätze 1926-1936*, **VII-XXXI**, Wien, Springer, 1938.

[11] 'Von der Natur eines philosophischen Problems,' *Synthese* 4, 340-350, 395-406, 1939. Reprinted in: [31], 81-103.

[12] 'Was ist logische Analyse?' *The Journal of Unified Science: Erkenntnis* 9, 265-289, Den Haag-Chicago, 1939/40. Reprinted in: [31], 42-66. English translation: 'What is Logical Analysis?' (in this volume).

[13] 'Verifiability', *Proceedings of the Aristotelian Society,* Supp. Vol. XIX, 119-150, 1945. Reprinted in: *Logic and Language* (First Series) (ed. by. A. Flew), 117-144, Oxford, Blackwell, 1951; and in *The Theory of Meaning* (ed. by G. H. R. Parkingson), 35-60, London, OUP, 1968, Partly translated into German in: *Sprache und Analysis* (ed. and transl. by R. Bubner) 154-169, Göttingen, Vandenhoeck & Ruprecht, 1968.

[14] 'Are there Alternative Logics?' *Proceedings of the Aristotelian Society* **XLVI**, 77-104, 1945/46. Reprinted in: [30], 67-90.

[15] 'The Many-level-structure of Language,' *Synthese* 5, 211-219, 1946. Reprinted in: [30], 91-101.

[16] 'Language Strata,' *Logic and Language* (Second Series), (ed. by A. Flew), 11-31, Oxford, Blackwell, 1953. Reprinted in: [30], 102-121.

[17] 'Logische und psychologische Sprachbetrachtung,' *Synthese* 6 460-475, 1947. Reprinted in: [31], 104-120.

[18] 'Analytic-Synthetic,' *Analysis* **10.2,** 25-40, 1949, **11.2** 25-38, 1950; **11.3,** 49-61, 1951; **11.6** 115-124, 1951; **13.1,** 1-14, 1952; **13.4.** 73-89, 1952. Reprinted in: [30], 122-207.

[19] 'Fiction' (1950), first published in this volume.

[20] 'A Note on Existence' (1952), first published in this volume.

[21] 'A Remark on Experience' (1950's), first published in this volume.

[22] 'The Linguistic Technique' (after 1953), first published in this volume.

[23] 'How I See Philosophy,' *Contemporary British Philosophy,* III, (ed. by H. D. Lewis), 447-490, London, George Allen & Unwin, 1956. Reprinted in: *Logical Positivism* (ed. by A. J. Ayer), Glencoe, The Free Press, 1959: [30], 1-38. German translation: 'Wie ich Philosophie sehe', [31], 121-163.

[24] 'Belief and Knowledge' (1950's), first published in this volume.

[25] 'Two Accounts of Knowing' (1950's), first published in this volume.

[26] 'The Decline and Fall of Causality', *Turning Points in Physics* (ed. by A. C. Crombie), 84-154, Amsterdam, 1959. Reprinted in: [30], 208-256.

[27] *The Principles of Linguistic Philosophy* (ed. by R. Harré), London-Melbourne-Toronto, Macmillan, 1965. [This is substantially an English translation of [32]. Italian translation: *I principi della filosofia linguistica,* (transl. by E. Mistretta), Roma, 1969.

[28] *Wittgenstein und der Wiener Kreis* (ed. by Brian McGuinness), Oxford, Blackwell, 1967; and as: Ludwig Wittgenstein, *Schriften,* Band 3, Frankfurt/M., Suhrkamp, 1967.

[29] 'Suchen und Finden in der Mathematik', *Kursbuch* **8**, 74-92, Frankfurt/M., Suhrkamp, 1967. Also in [32].

[30] *How I See Philosophy* (ed. by R. Harré), London-Melbourne-Toronto, Macmillan, 1968. [Contains: [13], [14], [18], [23], [26]].

[31] *Was ist logische Analyse?* (ed. by Gerd. H. Reitzig), Frankfurt/M., Athenäum, 1973. [Contains: [1], [3], [4], [8], [9], [11], [12], [17], [23]].

[32] *Logik, Sprache, Philosophie,* (ed. by G. P. Baker, Brian McGuinness, Joachim Schulte; Vorrede by Moritz Schlick; Nachwort by G. P. Baker and Brian McGuinness), Stuttgart, Reclam, 1976.

[33] *Philosophical Papers* (ed. by Brian McGuinness; introduction by Anthony Quinton), *Vienna Circle Collection,* Vol. 7, Dordrecht and Boston, D. Reidel, 1976. [Contains: [1], [3], [4], [6], [7], [8], [12], [19], [20], [21], [22], [24], [25].]

INDEX OF NAMES

VIENNA CIRCLE COLLECTION